Legal and Scientific Uncertainties of Weather Modification

Edited by
William A. Thomas
American Bar Foundation

For the
National Conference of Lawyers and Scientists,
a joint organization of
the American Bar Association and
the American Association for the Advancement of Science

**Proceedings of a symposium convened at
Duke University, March 11-12, 1976,
by the National Conference
of Lawyers and Scientists.**

Duke University Press
Durham, N.C.
1977

© 1977, Duke University Press

L.C.C. card no. 77-82058

I.S.B.N. 0-8223-0393-0

Printed in the United States of America

Contents

83-5721

Preface

The American Bar Association (ABA) and the American Association for the Advancement of Science (AAAS) in 1974 jointly created a multidisciplinary entity consisting of seven representatives from each organization.* This group, formally designated the National Conference of Lawyers and Scientists, adopted the objective of facilitating cooperation and communication among members of the two professions, as broadly defined, to help resolve important social issues and preclude future problems.

As its first major project, the group convened a conference to focus on the uncertainties surrounding current and foreseeable attempts to modify weather. This subject was selected because it encompasses a broad range of legal and scientific uncertainties, because it promises to become of increasing interest, and because it is not yet so emotionally charged that objective assessments are unattainable. The ABA, the AAAS, and Duke University sponsored the Conference on Legal and Scientific Uncertainties of Weather Modification on March 11-12, 1976, with assistance from the American Bar Foundation and the American Meteorological Society.

*The following persons served as representatives at the time of the conference:

Representing AAAS: Emilio Q. Daddario (Co-Chairman), Office of Technology Assessment, U.S. Congress; Robert Berliner, School of Medicine, Yale University; William Bevan, Department of Psychology, Duke University; Richard H. Bolt, of Bolt, Beranek, & Newman, Inc., Cambridge, Mass.; Ruth M. Davis, Institute for Computer Sciences and Technology, National Bureau of Standards; Vincent P. Dole, Rockefeller University Hospital, New York City; and David J. Rose, Department of Nuclear Engineering, Massachusetts Institute of Technology.

Representing ABA: W. Brown Morton, Jr. (Co-Chairman), of Morton, Bernard, Brown, Roberts & Sutherland, Washington, D.C.; Roger P. Hansen, of Hansen & O'Connor, Denver; Haywood H. Hillyer, Jr., of Milling, Benson, Woodward, Hillyer & Pierson, New Orleans; Harold Horvitz, of Guterman, Horvitz, Rubin & Rudman, Boston; Milton Katz, Harvard Law School; Lee Loevinger, of Hogan & Hartson, Washington, D.C.; and Preble Stolz, California Office of Planning and Research, Sacramento.

The objective of this two-day conference was to increase the availability of scientifically valid and legally acceptable advice on weather modification to administrators, legislators, and judges, thus promoting sound decision making and public confidence in how the legal system handles scientific issues. Placing emphasis on the legal and scientific uncertainties associated with weather modification afforded an excellent opportunity to discuss multidisciplinary problems, and the hospitable setting at Duke University and the program format encouraged open and candid discussion. It was my pleasure to serve as moderator. All persons cited herein approved publication of their edited remarks and are further identified in the list of participants at the end of this volume.

The ABA and the AAAS did not intend to resolve substantive legal or scientific issues at this conference. Instead, the subject was chosen as a paradigm of the need for stating issues carefully to determine with more confidence the current situation and ways of improving it. The deliberations of the three working groups attest to this necessity. They each met twice during the conference to discuss current issues that require attention, probable future issues that might be avoided by proper planning and action, and priorities for allocating resources to solve or preclude problems.

The 80 invited participants included private practitioners in both science and law; staff members of legislative reference services, of scientific organizations, and of private and public research organizations; members of the federal judiciary; and academicians of various disciplines. Scientists outnumbered lawyers but not disproportionately relative to the types of topics under consideration.

This certainly is not the first attempt to discuss legal and scientific issues related to weather modification, as indicated by the literature citations throughout the volume. As one example, the National Moot Court competition in 1950, sponsored by the Association of the Bar of the City of New York, concerned the legal issues raised by a weather modification project. Representatives of 42 law schools argued a hypothetical case involving a contract to seed clouds, the success of which allegedly deprived a downwind rancher of needed precipitation. The trial court had the benefit of expert scientific witnesses in helping to understand the meteorological phenomena and in assessing the monetary damages.

Several points of consensus evolved during the conference:

First, there will be increased attempts to modify weather, both because people tend to do what is technically possible and because the anticipated benefits of precipitation augmentation, hail or lightning suppression, hurricane diversion, and other activities often exceed the associated costs.

Second, the central issue, as with many developing technologies, appears to be whether we should employ prospective safeguards or rely solely upon retroactive measures. We are unable to predict accurately the secondary and possibly tertiary physical and social effects of weather modification. Reliance on statistical assessments of cause and effect raises legal issues as yet unresolved.

Third, weather modification should be considered a means toward achieving national or international goals—*e.g.,* ecosystem stabilization and food production—and not an end in itself.

Fourth, scientists and lawyers appear to be more inclined to cooperate on policy matters than they were in the past. Scientists often are reluctant to participate within the legal system because they believe that technical issues should be resolved by the scientific community. Unfortunately, law cannot always wait for the certainty of science, and persons working within the legal system often must determine what constitutes a "fact" without the benefit of an extended investigation. Members of both professions understand that scientific knowledge accumulates over time and that scientific "facts" are more likely than not subject to change. Despite increasing discussion about the need to modify the existing "fact-finding" methods of the legal system to accommodate technological issues, the attributes of the legal and the scientific methods of "fact finding" are not incompatible in addressing these issues.

While the need to focus on specific subjects of mutual concern tended to discourage broad policy discussion, as is almost inevitable in new multidisciplinary efforts, trends toward increasing cooperation and communication were evident at the conference. For example, although it was not difficult to distinguish the lawyers from the scientists during the morning of the first day, it became increasingly difficult to do so as the scientists interpreted legal issues and the lawyers commented on resolution of scientific and technical problems. The serious discussions were interspersed with lighter moments, as when David Rose suggested that the conference adopt as its fight song, "Gang, Gang, the Hail's All Here."

The sponsors and most of the participants considered the conference a successful first step toward achieving its objective. Both *Science* and the *American Bar Association Journal* published a summary of the conference, and the conference has been cited in congressional testimony. In addition, the National Conference of Lawyers and Scientists, the American Meteorological Society, and the Atmospheric and Hydrological Sciences Section of AAAS are cosponsors of a symposium on the technical and legal aspects of weather modification at the AAAS annual meeting in 1977.

Neither the sponsors of this conference nor the authors consider this volume to be a definitive statement. Rather, they view it as a status report on how relevant questions are being addressed and on the need to design more effective multidisciplinary arrangements. Representatives of the parent organizations solicit comments on future projects of social importance. These proceedings are published with the hope of encouraging thoughtful persons to consider the pressing problems that involve both legal and scientific uncertainties and to work toward overcoming the institutional and social inertia we now face.

William A. Thomas
American Bar Foundation
Chicago, Illinois 60637

Welcoming Comments
Frederic N. Cleaveland, Provost

Duke University
Durham, North Carolina 27706

It is a pleasure to welcome this group to Duke University, first because of your worthy objectives, and second because I recognize among you so many whom I have met before on earlier visits to the campus. We are very pleased that the organizers of this conference chose the Duke Law School as the setting for your deliberations.

I extend to you a warm welcome from President Sanford and from the entire Duke community and assure you a hospitable reception during these two days.

Duke University is very much committed to the purpose of this conference. We see in your meeting another important opportunity to build more effective bridges linking science, law, and public policy. You can measure our commitment to this purpose by observing some of our programs, such as the Round Table on Science and Public Policy that Professor William Bevan and Dean John McKinney have conducted over the past several years. This program brings to the campus leading contemporary statesmen of science, some of whom are in the audience today, in an effort toward building these bridges.

The program of the Institute of Policy Sciences and Public Affairs and the activities of the Duke law faculty and students and others within the University also reflect these overall purposes. As an institution we are engaged in working continually to promote the objectives of your conference. Finally, we take pride in the long-standing and close-working relationships Duke faculty members have maintained with both parent organizations—the American Bar Association and the American Association for the Advancement of Science. For that reason it is an even greater privilege for us to have you convene this conference here.

You have scheduled two busy days. Only a casual look at the agenda is required to conclude that this is a working conference on an important social

1

issue; it is not one of those affairs to be filled with rhetoric for the sake of rhetoric. With that thought in mind, I'll not take more of your time but let you get on with the main business of the day.

Introduction
Emilio Q. Daddario, Director

Office of Technology Assessment
Congress of the United States
Washington, D.C. 20510

I am pleased that we are beginning to consider a serious problem of the future and to conduct an experiment we hope will prove of great value to our society.

We are here because we recognize the reality of social problems—that they increasingly will be attributed to factors involving the application of scientific knowledge. On that score we cannot assume that there will be neither a lesser need nor a lesser desire on the part of society to acquire that knowledge. Therefore, we must accept the responsibility of searching for ways to put new knowledge and new technology to use for the overall benefit of society within the structure of our legal system. We do this to influence the decision-making process in both the private and the public sectors.

We must acquire in this process greater wisdom about how science and technology are applied and about how they emerge as issues that excite public awareness and emotions to the point where it sometimes is difficult to appraise them dispassionately. Even more important, we must find ways to assess objectively the individual decisions that lead to the development of the overall problem. We should learn through examination of that process how we best can help apply knowledge toward solving social problems. For this reason, the American Bar Association and the American Association for the Advancement of Science decided to convene this conference to learn how better communications between the scientific and legal communities could be attained.

We cannot assume, as we often have in the past, that we automatically will choose the right course just because we have a democratic form of government. Our decision-making processes are fragile, and much work is needed to develop institutions that make our democracy work. Although our society is a fragile one, it is the type that best serves the individual and the general public.

We must develop more effective communications among lawyers and scientists to see that this continues. I appreciate the opportunity on behalf of the AAAS to welcome the distinguished participants to this conference and look forward to these two days with much excitement and expectation about what we can accomplish.

Introduction

W. Brown Morton, Jr.*

Morton, Bernard, Brown, Roberts & Sutherland
Washington, D.C. 20006

I welcome you here on behalf of the American Bar Association. It occurred to me that all of us—both lawyers and scientists—might like to know something about the organization responsible for this conference. The American Bar Association has a variety of national conference groups that are composed of members of the ABA and members or representatives of other professional disciplines. The conference that Mr. Daddario and I co-chair is known as the National Conference of Lawyers and Scientists. Some of the other National Conferences involve lawyers and bankers, lawyers and engineers, lawyers and social workers, for example. The objective of our conference, as I see it, is to keep lawyers and scientists from being in positions counter to one another. The officers of the ABA realize the importance of the interplay between science and law and the need in the foreseeable future for increased cooperation. They are very eager for this conference to succeed in enabling law to work better with science and to allay some of the difficulties scientists have encountered in working with lawyers. I am particularly sensitive to this because I am engaged principally in the trial of patent cases.

A March 10, 1976, editorial in the *Washington Star*, "Judgment and Science," reminds us: "The study of science no more guarantees detachment than the study of law guarantees judiciousness." I think all of us should keep that well in mind.

Now, since we are not here to examine the role of science versus that of law but of how science works with law, we should remember the importance of rendering unto law those things that truly are the law's and of rendering unto science those things that truly are science's. In the familiar legal dichot-

*Current address: Morton & Roberts, 1800 M St., NW, Washington, D.C. 20036.

omy, I look to scientists for knowledge to allow the law to make "findings of fact" and to lawyers for judgments on proper conclusions of law. The law is surely brought into disrepute when it proceeds on scientific falsity, as it has done sometimes by accepting scientific anachronisms, but the law is not always prepared to rid itself of some cobwebs.

The law must regulate the activities of science under the same standards of due process with which it attempts to regulate the activities of other forces in society. Thus, scientists must tolerate some aspects of the law that may seem cumbersome but that really are essential to the operation of our system of due process. Lawyers in turn must recognize that perhaps we must modify the cherished adversary system and evolve some new procedures for dealing with particular scientific matters that are essential if courts are to render sound judgments.

Scientific and Other Uncertainties of Weather Modification

Lewis O. Grant

Department of Atmospheric Science
Colorado State University
Fort Collins, Colorado 80523

Introduction

There is a wide diversity of opinion on weather modification. Some believe that weather modification is now ready for widespread application. In strong contrast, others hold that application of the technology may never be possible or practical on any substantial scale. Participants in this conference have views covering the full range of views on the subject, and many of these views will be presented here.

Let me begin by stating unequivocally that at least some clouds found in the atmosphere can be modified and that this modification will result in changes in the amount of precipitation. For example, supercooled clouds (liquid-water clouds with temperatures below freezing) can be converted to ice-crystal clouds by artificial treatment (cloud seeding). The conversion is accomplished by supplying artificial nuclei for the formation of ice crystals to compensate for a deficit in natural ice nuclei in the atmosphere. The ice crystals can grow rapidly at the expense of small, subcooled, liquid cloud droplets to form precipitation particles large enough to fall from the cloud.

This clear example of intentional weather modification has been repeatedly demonstrated. Sometimes the precipitation at the ground will be greater from this type of cloud modification, but at other times it can be less. It has also been demonstrated that some clouds can be induced to grow larger by seeding. It also is well known that weather—*e.g.,* rain, hail, and thunderstorms—is sometimes affected by the presence of large cities. These are a few of the findings from 30 years of work since the discoveries of Langmuir and Schaefer in 1946, which provided a scientific basis for weather modification.[1]

Present capabilities are too limited to justify extensive application. Scientific advances have been sufficient to conclude that the probability is high that substantially broader atmospheric modifications are scientifically and technologically feasible. It is likely that progress in weather modification can accelerate, now that the problems are becoming better defined and the tools and methods for dealing with them are becoming available.

The capabilities of weather modification must be kept in careful perspective. Climate is controlled primarily by the earth's incoming radiation, the constituents of the earth's atmosphere, and the atmospheric interaction with land and water masses. Although we ultimately may modify these—and may now be doing so inadvertently—the controlled modification of these primary climate controls is not yet within reach. Weather changes that now can be considered involve slight alterations of precipitation and radiation and, possibly, slight reductions in some catastrophic events, such as hail and lightning. Perhaps during the next quarter-century some appraisal will be possible of large-scale alterations of weather and climate. Meanwhile, the emphasis of this presentation is on the slight changes in weather that may be feasible now or in the near future.

I discuss first the terminology we will be using at this conference. There are three key words: "law," "weather modification," and "uncertainty." I will comment on "weather modification" and "uncertainty," while leaving "law" to the lawyers.

"Weather modification" is an extremely broad term. I am frequently asked, "Does weather modification work?" but I am not really sure what I have been asked. The answer is clearly "Yes," since cold clouds or fog can be cleared from airports and since some orographic (mountain) clouds can be seeded to increase precipitation. The "yes," however, is misleading since it has not been satisfactorily demonstrated that summertime rainfall can be increased over an area on a determinate basis, that hail can be decreased, that severe weather can be abated, or that lightning can be suppressed. When a concerned farmer in a drought area asks, "Does weather modification work?" he probably really means, "Can weather modification be used to provide some water for my crops?" Too often the broad "yes" to the question on the state of the art of weather modification is misinterpreted and really does not address the intended question.

"Uncertainty" also warrants a brief comment. Uncertainty may imply "falling just short of certainty" or it may imply "an almost complete lack of definite knowledge." Weather modification in its various forms falls within both categories, and we must keep this in mind throughout the conference. In certain instances, we may use uncertainty to mean "almost certain," while in other cases we may mean "almost completely lacking in definite knowledge." Perhaps this conference for improving communication between law and science will be a start toward providing more precise terminology.

Possibilities and Capabilities

What Is Possible or Impossible?

Some modifications of weather clearly are possible and practical. Other desired modifications clearly are impossible.

Possible and clearly practical cloud modifications include supercooled fog dissipation over airports and precipitation augmentation from orographic clouds under certain conditions. It is also possible, but perhaps not frequently practical, to augment precipitation from single, small cumulus clouds. Certain other types of cloud modification appear feasible but have not been consistently and clearly demonstrated—e.g., summertime precipitation augmentation over a large area and hail and lightning suppression. The uncertainties about still other suggested applications of weather modification are so great that they cannot even be evaluated reasonably at present. These include decreases in radiation by the creation of high-level clouds over large areas, changes in characteristics of meso- and large-scale storm systems, and others.

Some impossibilities can be addressed briefly, but they are obvious or have little meaning. It is, for example, impossible to create rain from a clear atmosphere with low humidity. It is impossible to create rain, as opposed to snow, when temperatures throughout the atmosphere are below freezing. It seems impossible to consider the alleviation of drought over large areas, the elimination of floods, or the moderation of cold air masses that cause widespread frost damage. For all practical purposes at this conference, we clearly can treat these modifications of the weather as being impossible. It is, however, not certain that incoming radiation cannot be affected over large areas or that the energy distribution within general storms cannot be modified selectively. Such alterations might affect the patterns, movements, characteristics, or intensities of meso- and large-scale storms, which in turn are the determinants of droughts and floods.

At present, weather modification can be designated either as clearly possible and practical or as clearly impossible under a very limited set of conditions. The variability of natural atmospheric processes is sufficiently large and the effects of slight alterations are so unclear that the possibility of weather modification on most weather systems cannot be evaluated adequately. We should focus our legal considerations only on the limited types of weather modifications that now are possible, but the broader possibilities with wide-scale implications should not be ignored. Whether real or perceived, these may well lead to the greatest legal and political conflicts concerning application of weather modification.

What Is the Current Status?

It seems useful to consider the current status of various phases of weather modification.

1. Fog

A usable technology now exists for clearing some cold, supercooled fogs from specific areas at relatively low cost. Capabilities for clearing warm fogs (temperature greater than $0°$ C) are limited. The most effective method, massive local heating, is very expensive.

2. Orographic clouds

A practical technology now exists for augmenting precipitation from some wintertime orographic clouds. Increases in overall precipitation on the order of 5-20%, depending primarily upon the location, can be expected with some confidence. Capabilities are likely—but have not been demonstrated consistently—for some types of summertime orographic clouds.

3. Convective clouds (clouds with substantial vertical atmospheric motions)

There is good evidence that seeding of wintertime cloud systems with imbedded convective elements might increase precipitation by 10-20% under certain conditions.

While changes in precipitation from cloud seeding have been demonstrated for certain convective summertime clouds, augmentation of precipitation over a large area has not been demonstrated with a satisfactory degree of certainty. Experimental evidence is conflicting. The results from at least four carefully conducted experiments show decreases in precipitation following seeding.[2] At least two recent experiments using more advanced concepts indicate precipitation increases, but still with a high degree of uncertainty.[3, 4]

Research seeding of convective clouds has also been conducted to reduce hail and to suppress lightning. Again, the evidence of success is conflicting. Two of the most carefully designed statistical hail experiments failed to produce the desired suppression effects. Other experiments in the United States and elsewhere provide evidence that seeding reduces hail damage to crops. The variable evidence from seeding convective storms to reduce severe weather attests to the moderate to high level of scientific uncertainty of weather modification to alleviate catastrophic effects.[5]

*4. General storms (mesoscale systems and tropical and
    extratropical storms)*

Most natural precipitation comes from organized cloud systems, or general storms. It is likely that any substantial changes in precipitation over large areas also would have to be associated with these systems. Although seeding operations have been conducted during these general storms, controlled field experiments that provide definitive results have not been performed.

5. Other types of weather modification

Numerous other possibilities for weather modification have been proposed. These include controlling either incoming or outgoing radiation to modify temperature extremes, reducing strong thunderstorm winds, and suppressing tornadoes. In general, specific proposals are not sufficiently advanced to provide even the basis for extensive field experimentation.

6. Extended area effects

There is increasing evidence, particularly from wintertime programs, that the effects of some weather modification experiments may extend over considerably greater areas than intended. Present evidence suggests that precipitation is frequently increased under some conditions for distances of 50 to 150 miles (80 to 240 km) from the intended targets. No significant evidence exists that precipitation decreases downwind of seeded areas. The degree of uncertainty about the reality and magnitude of downwind effects and associated physical processes is great.

7. Inadvertent weather modification

Evidence of inadvertent weather and climate modification resulting from human activity continues to accumulate. The atmospheric concentration of carbon dioxide has been increasing since the start of the Industrial Revolution. Its effects on the atmospheric radiation balance are well established, and temperatures in the lower atmosphere should increase with increases in carbon dioxide. This heating effect could cause many changes in weather and climate. In contrast, the increase in atmospheric particulates due to human activity can cause optical and chemical changes that could decrease atmospheric temperatures. The net effect of these opposing effects has not been established.

Furthermore, recent studies downwind of large cities provide substantial evidence of additional types of inadvertent weather modification, probably owing to the "heat island" effect of the city. Changnon and Semonin in 1974 reported increases of up to 20% in rainfall and, based on strong evidence, more frequent thunderstorms and a 100% increase in the frequency of hailstorms east of St. Louis.[6]

What Is the Present Technology?

Turning now from considerations of the current status of various types of weather modification, I will describe briefly the present technology used in the various phases in the seeding process.

1. Cloud-seeding materials

Excellent progress has been made in producing cloud-seeding nuclei (very small particles on which water in clouds collects) with a wide range of specified characteristics. In successful contemporary seeding efforts, when such nuclei are delivered to a cloud, their presence induces a change in the physical phase of water within the cloud and consequently produces a change in the cloud itself. To a substantial degree, seeding can be tailored to meet specific requirements for different clouds. Some seeding materials may remain in the atmosphere for a long time; other, biodegradable materials have a short life span.

2. Delivery and transport of seeding material

The delivery of seeding materials to the interior of clouds constitutes a particularly weak link in the application technology. Techniques have been developed for releasing the seeding materials from ground sites, from aircraft flying at cloud base or in clouds, and from rockets dropped from aircraft above the clouds. There are still a number of unresolved problems in ensuring the delivery of seeding materials in the proper concentrations at the right location at the proper time.

3. Determination of seeding potential

Knowledge of how a specific cloud system should be treated is a prerequisite for effective weather modification. For many cloud systems, such as cold fogs and some orographic clouds, this determination can be made, and we can specify simplified requirements that permit some applications on a determinate basis. For most types of clouds, much more complex descriptions are required, and moderate to highly sophisticated numerical computer models will be needed to integrate the many variables. These models now are becoming available.

For efficient application, even real-time definition of seeding potential is not sufficient. Lead time is needed to make the seeding treatment, and this requires predictor variables of atmospheric conditions that can be used to forecast seeding potential at least a few hours in advance. Again, except for cold fogs and some orographic clouds, capabilities are inadequate for most applications and for many research experiments.

4. Evaluation of results

The capability of specifying quantitatively the effects of individual seeding operations, or even those for extended time periods, is limited. It is seldom possible even after careful analysis to specify the results of cloud seeding on a case-by-case basis in an application program. The range of natural variability of clouds and rainfall is too great, and our understanding of natural processes and of seeding effects is incomplete. Interpretations of seeding effects can be handled only on a probabilistic basis from physical considerations and from comparisons with similar treatments in carefully conducted experiments. This uncertainty about what happens after individual weather modification treatments adds significantly to the legal complexities.

Substantial progress has been made, however, in evaluating the results of sustained cloud seeding. The design requirements for detecting changes in precipitation from as little as 10-20% superimposed on a natural variability of 250-400% are being clarified. Better descriptions of the natural differences among clouds and better comprehension of the questions that must be answered are important aspects of increasing this capability. Perhaps of greatest importance in evaluating weather modification results is the communication that has been established between well-trained researchers specializing in cloud physics and weather modification and a small and scattered group of competent statisticians with interest in the experimental design and evaluation of weather modification programs.

5. Observational capabilities

Development of instruments for observing natural and altered cloud processes has advanced significantly. Platforms for observing atmospheric nuclei, cloud water and ice particles, and cloud motions include gliders that can follow upward-moving air currents through clouds and aircraft specially armored to allow safe travel through hailstorms. These observational systems provide a new and vital basis for important advances that previously were impossible.

6. Societal and environmental considerations

The direct economic value of certain weather modification attempts seems clear, as discussed to some extent later in this paper. Farhar has described weather modification as "a collective innovative decision affecting entire communities or regions, in contrast to individual innovative decisions affecting primarily personal matters."[7] For some applications, the total economic gains appear to more than offset total costs, including social ones. The uncertainties, however, are great. Decisions for implementation can be reached through in-depth analyses of environmental, economic, social, legal, and political constrants.

The Key Uncertainties

What Are Natural Cloud Processes?

The key uncertainty in weather modification is the ambiguous nature of processes within natural clouds. After centuries of observation and several decades of intensive study of the atmosphere, we have only a modest understanding of how precipitation forms. It was less than 40 years ago that the importance of the growth of ice crystals from the depletion of coexisting water droplets was realized. Twenty-five years ago the theory of how precipitation forms by coalescence in clouds with a temperature at about freezing was just gaining acceptance. The inability to define the outcome of much cloud seeding does not result from a lack of knowledge about seeding materials, an inability to deliver these seeding agents, or any failure of these seeding materials to alter the seeded cloud. It results from a lack of understanding of the complex natural cloud processes and their interactions. As a result, we cannot say with certainty when and how the technology should be applied.

The physical differences among clouds, many of which are visually similar, are great and affect substantially the potential for weather modification and the choice of seeding techniques required. The lifetime of an orographic cloud is typically 7-8 hours, while that of a moderate-sized cumulus cloud cell is typically one-third of an hour. The cumulus system may be short-lived or may last for a number of hours. The vertical velocities of orographic and cumulus clouds are typically 0.33 feet a second (0.1 meters/second) and 15-60 feet a second (5-20 meters/second), respectively. The number of atmospheric nuclei necessary for the formation of cloud droplets and of cloud ice particles also varies and greatly affects the efficiencies of clouds to produce rain or snow. All of these variations and their effects on the precipitation processes are poorly defined. Because we do not understand how natural processes operate, we are unable to determine when opportunities exist to increase precipitation. To reiterate, the lack of understanding of natural cloud processes constitutes the key uncertainty of weather modification.

Can General Storms Be Modified Beneficially?

Uncertain knowledge of atmospheric processes at the scale of general storms is also large. Since most natural precipitation is produced by these systems, it seems clear that they must be modified if weather modification is going to have much effect. Gaps in knowledge relate both to precipitation efficiencies of existing storm systems and to the change in dynamics that could alter the amount of atmospheric water produced by them. These systems are even more complex than local convective clouds, so that description of the natural systems and of the methods for altering them are even less refined. The uncertainties preclude reasonable consideration of the weather modification potential of these meso- and large-scale storms.

Do Extra-Area Effects Exist?

A third major uncertainty concerns the geographic extent of weather alterations. Unintentional effects beyond the intentional seeding areas can affect many more persons and communities than those for which the activity was planned. Understanding these extra-area effects may help in the planning for large-scale treatments of extensive areas. Evidence for broad area effects from local weather modification efforts is increasing, but it is difficult and expensive to verify. At least partial resolution of this uncertainty will be required before many types of weather modification experiments and operations can be conducted and will be needed to resolve questions of societal concern, including many legal issues.

How Serious Are Societal Implications?

A fourth major uncertainty specifically involves issues related to the overall effect of weather modification on society: economic benefits and risks, environmental effects from weather modification, and effects on societal standards and institutions, including legal and political structures. The complexity of these issues increases in the more populated areas with more diverse activities. The magnitude of these societal impacts and the methods for resolving them raise uncertainties that also must be resolved before substantial implementation, or even some key research experiments, can be conducted.

Priority Needs

Four of the priority needs for resolving the key uncertainties of weather modification clearly can be identified.

Commitment and Policy

Commitment by government and a well-defined policy will be required for reasonably rapid and steady progress in weather modification. While progress will continue from present efforts, current support levels are insufficient to address many of the unknowns that must be resolved to develop a broadly applicable technology in the near future. The policy adopted should include institutional arrangements for overall planning, organization, and management of programs that place emphasis on scientific and societal problems of high priority.

Field Experiments

Well-organized and adequately funded field experiments in several geographical areas for periods of at least five to ten years will be needed to resolve the key uncertainties of natural and intentionally altered cloud processes. The complexity of atmospheric processes and of their interactions precludes resolution of most key questions by laboratory experiments. Field experiments should emphasize systematic observations on size scales ranging from angstroms (nucleation of liquid and ice particles), to millimeters (precipitation), to hundreds of meters (small clouds and cloud depths), to thousands of meters (large-scale storms).

Basic Research

Basic research to provide a broader scientific base is an additional priority need for adequately considering potential weather modification opportunities. It seems likely that man-made changes in atmospheric composition, radiation, and other characteristics can have more far-reaching effects than those now contemplated through cloud seeding. These effects seem to be occurring inadvertently already.

Resolution of Societal Impacts

Resolution of the societal impacts is an additional concern. The overall issue of risks and benefits surrounds both research and implementation, and this issue already imposes serious constraints on much needed research aimed at defining the problems and effects. This priority applies to all societal aspects, including economic, environmental, social, legal, and political, and others to be discussed at this conference.

Is Weather Modification Worth the Effort?

The answer clearly is "Yes." The effects can be large and the prospects of success are high. It is widely recognized that some potential already exists. It appears that snowpack increases of 5-15%, for example, are feasible with present techniques. Increases of this magnitude would add around 10 million acre-feet (12.2 billion cubic meters) of water to irrigation systems in the western United States. Even if it enters the water-use priority system at the lowest end of the value structure (lower value agricultural crops), the value of the additional water would be several hundred million dollars a year. When one considers the reuse of irrigation water and the higher value of still other uses, the direct gains in production at 1975 prices would probably be in excess of $1 billion a year. The direct cost of the weather modification application would not be high, and considerable margin would still exist in the cost/benefit analyses to apply to other costs, such as indirect and social ones and compensation for those adversely affected by the program.

While the state of readiness of other types of weather modification is less certain, the gain in agricultural production would be great with even modest changes in precipitation. The value of annual yield increases in just the main production areas for wheat, corn, soybeans, and range forage as a result of a 10% increase in precipitation would exceed $2 billion at 1975 prices. As strains on food and energy supplies increase, this additional production will become vital.

The potential for some relief from catastrophic events—*e.g.,* hail, hurricanes, and lightning—is real. Such relief could affect individual lives, society, and the nation's economy.

Federal funding for weather modification research in 1974 equaled an expenditure of only about 0.7% of the gain in agricultural production that would result yearly from a 10% increase in precipitation and a 30% decrease in hail damage for just the key crops in the main production areas of the United States. Stated another way, if weather modification research continued at present levels and constant dollars for 150 years and accomplished what is now deemed feasible or reasonably possible, the increased agricultural product in any one year then would offset the cost of all 150 years of research. The annual tax revenues (as distinct from the value of the increased production) from the increased production would be at least 10 times greater than present annual research expenditures.

When one considers that the gains can be large and that the prospects of success are high (almost certain for orographic precipitation augmentation), it becomes clear that weather modification merits a substantially greater commitment.

Summary

Important and steady advances have been made in developing technology for applied weather modification, but complexity of the problems and lack of adequate research resources and commitment retard progress. Advances have been made in training the needed specialists, in describing the natural and treated cloud systems, and in developing methodology and tools for the necessary research. Nonetheless, substantial further efforts are required.

The key remaining uncertainties are:

1. The ambiguous nature of natural and seeded cloud processes.
2. The incomplete understanding of the feasibility for useful modification of meso- and large-scale storm systems.
3. The insufficiency of data on the areal and temporal extent of weather modification effects.
4. The inadequate understanding of the societal influences of weather modification.

The key uncertainties can be resolved by:

1. Commitment to develop and implement a sound national policy to support weather modification research.
2. Comprehensive field experiments in several geographic areas to provide missing information on cloud processes and weather systems and to develop and test appropriate treatment technologies.
3. Increased basic research to encourage innovative approaches.
4. Parallel and systematic attempts to resolve the societal issues.

Notes

1. V. J. Schaefer. 1946. The production of ice crystals in a cloud of supercooled water droplets. Science 104:457-59.

2. L. O. Grant, G. W. Brier, & P. W. Mielke. 1974. Cloud seeding effectiveness for augmenting precipitation from continental convective clouds. *In* Proc., International Tropical Meteorological Meeting, Nairobi, Kenya. American Meteorological Society.

3. S. A. Dennis, P. L. Davis, H. J. Hersch, D. E. Cain, & A. Koscielski. 1974. Cloud seeding to enhance summer rainfall in the northern plains. Institute of Atmospheric Science, Rapid City, S. D. 161 p.

4. W. Woodley, G. Sambataro, J. Simpson, & R. Biodini. 1976. Rainfall results of the Florida area cumulus experiment, 1970-1975. Proc. Second World Meteorological Organization Conf. on Weather Modification. International Association of Meteorology and Atmospheric Physics and American Meteorological Society.

5. National Center for Atmospheric Research. 1976. Report on Hail Suppression Symp. National Center for Atmospheric Research, Boulder, Colo.

6. S. A. Changnon & R. G. Semonin. 1974. Results from Metromex, 1970-1971. Bull. Am. Meteorol. Soc. 55:171-78.

7. B. C. Farhar. 1975. Weather modification in the United States: A sociopolitical analysis. Ph.D. Thesis, Univ. Colorado, Boulder.

General References

Battan, L. J., & A. R. Rassander, Jr. 1967. Summary results of a randomized cloud seeding project in Arizona. Proc. Fifth Berkeley Symp. Math. Statist. & Probability. 5:29-34.

Bergeron, T. 1959. The problem of an artificial control of rainfall on the globe: General effects of ice-nuclei in clouds. Tellus 1:32-50.

Boone, L. M. 1974. Estimating crop losses due to hail. Agr. Econ. Rep. No. 267. U.S. Dep. Agriculture, Washington, D.C.

Braham, R. R., Jr., & P. Squires. 1974. Cloud physics—1974. Bull. Am. Meteor. Soc. 55:543-56.

Brown, K. J., & R. D. Elliott. 1971. Large scale effects of cloud seeding. Aerometrics Res., Inc. Rep. No. ARI-71-1. 42 p.

Changnon, S. A., Jr. 1975. The paradox of planned weather modification. Bull. Am. Meteor. Soc. 56:27-37.

Changnon, S. A., Jr., & F. A. Huff. 1971. Evaluation of potential benefits of weather modification on agriculture. Ill. State Water Surv. Rep. Part I:54-73.

Chappel, C. F. 1970. Modification of orographic clouds. Atmos. Sci. Paper No. 173. Atmospheric Science Dep. Colorado State Univ., Fort Collins. 196 p.

Chappel, C. F., L. O. Grant, & P. W. Mielke. 1971. Cloud seeding effects on precipitation intensity and duration of wintertime orographic clouds. J. Appl. Meteorol. 10:1006-1010.

Dennis, A. J., J. R. Miller, D. E. Cain, & R. L. Schwaller. 1975. Evolution by Monte Carlo tests of effects of cloud seeding on growing season rainfall in North Dakota. J. Appl. Meteorol. 5:959-69.

Farhar, B. C. 1976. The state of the art of weather modification: A survey of weather modification experts. J. Weather Modification (in press).

Fleagle, R. G., J. A. Crutchfield, R. W. Johnson, & M. F. Abdo. 1974. Weather modification: Science and research policy. Univ. Washington Press, Seattle.

Grant, L. O., & A. M. Kahan. 1974. Weather modification for augmenting orographic precipitation, p. 282-317. In W. N. Hess [ed.], Weather and climate modification. John Wiley & Sons, New York.

Grant, L. O., & J. D. Reid. 1975. Workshop for an assessment of present and potential role of weather modification in agricultural production. Atmos. Sci. Paper No. 236. Dep. Atmospheric Science, Colorado State Univ., Fort Collins.

Hess, W. N. [ed.]. 1974. Weather and climate modification. John Wiley & Sons, New York. 842 p.

Hosler, C. L. 1974. Overt weather modification. Rev. Geophys. Space Phys. 12:523-527.

Huff, S. A., & R. G. Semonin. 1975. Potential of precipitation modification in moderate to severe droughts. J. Appl. Meteorol. 14:974-79.

Juisto, J. 1974. Weather modification outlook—1985 projection. J. Weather Modification 6:1-13.

McQuigg, J. D. 1975. Economic impacts of weather variability. Compendium of sources on relationship between weather and man's activities. Atmospheric Science Dep., Univ. Missouri, Columbia.

Mielke, P. W., L. O. Grant, & C. F. Chappell. 1971. An independent replication of the Climax wintertime orographic cloud seeding experiments. J. Appl. Meteorol. 10:1198-1212.

Miller, J. R., E. L. Boyd, R. A. Schleusener, & A. J. Dennis. 1975. Hail suppression data from western North Dakota, 1969-1972. J. Appl. Meteorol. 14:755-62.

National Academy of Sciences. 1973. Weather & climate modification: Problems and progress. Washington, D.C. 258 p.

Neyman, J., E. I. Scott, & M. A. Wells. 1973. Downwind and upwind effects in the Arizona cloud seeding experiment. Proc. Natl. Acad. Sci. 70:357-60.

Rudel, R. D., H. J. Stockwell, & R. G. Walsh. 1973. Weather modification: An economic alternative for augmenting water supplies. Water Resources Bull. 9:116-28.

Sax, R. I., S. A. Changnon, L. O. Grant, W. F. Hitschfeld, P. V. Hobbs, A. M. Kahan, & J. Smith. 1975. Weather modification: Where are we now and where should we be going? An editorial overview. J. Appl. Meteorol. 14:652-72.

Schickendanz, P. T., & S. A. Changnon. 1975. Analysis of crop damage in the South African hail suppression efforts. Ill. State Water Surv. Rep. 27 p.

Simpson, J., W. L. Woodley, A. H. Miller, & C. F. Cotton. 1971. Precipitation results of two randomized pyrotechnic cumulus experiments. J. Appl. Meteorol. 10:526-44.

Singer, S. F. [ed.]. 1975. The changing global environment. D. Reidel Publishing Co., Dordrecht, Holland.

Weisbecher, L. W. 1972. Technical assessment of winter orographic snowpack augmentation in the upper Colorado River Basin. Stanford Research Institute, Menlo Park.

White, F. W., & J. E. Haas. 1975. Assessment of research on natural hazards. Univ. Colorado Press.

The Scientific Uncertainties:
A Lawyer Responds

Roger P. Hansen*

Hansen & O'Connor
American National Bank Building
Denver, Colorado 80202

I find it remarkable that the technology of weather modification has advanced so little after 30 years of research, experimentation, and operational field testing. We still do not understand the complexities of weather systems and climatic regimes, and the literature is rife with such terms as "tentative," "unknown," "not yet defined," and "some evidence suggests." It is very difficult to get anyone in the weather modification field to make a definitive statement. This, however, is laudatory in light of the uncertainties involved. The major areas of weather modification—precipitation augmentation, hail suppression, lightning suppression, and hurricane modification—include only two that seem ready for application: (a) wintertime augmentation from orographic clouds to increase snowpack in places like southwest Colorado and (b) cold fog clearance at airports.

The major efforts in weather modification are undertaken to increase agricultural production of food and fiber. These efforts obviously have vast potential for economic and humanitarian benefits, but it is amazing that so little has been written about other possible benefits, which of course have their own disadvantages. There obviously is a conflict, for example, between the golfer and the cotton grower in the Phoenix area. Recreation could be a limited objective of modification efforts, such as increasing snowpack in ski areas and providing more comfortable atmospheric conditions (objectives to which we have given little attention). Other possibilities include improvement of wildlife habitat—forage for deer or water for alligators—and provision of water resources for hydroelectric power or new energy developments, like oil shale development and coal gasification and liquefaction, which require large

*Current address: Industrial Environmental Research Laboratory, U.S. Environmental Protection Agency, Research Triangle Park, NC 27711

quantities of water. But the major purpose has been to increase agricultural production.

There is always a danger in being oriented toward a single goal. An obvious example is the desire to kill bugs on tomato plants without worrying about other possible consequences. If we look only at one goal, we easily ignore the possible synergistic effects and countless interrelationships. For instance, increased agricultural production would benefit consumers but could, at the same time, cause economically undesirable effects. Would it be desirable, for example, to use weather modification as a sort of price control device for agricultural products?

Most scientists seem to agree that we know even less about the socioeconomic and biophysical effects. Here, again, we are witnessing a societal lag behind technology. We have at best only started to investigate the political, legal, social, and secondary economic effects of weather modification.

I hear two schools of scientific and political thought. The first contends that the time is ripe for widespread application of what little we do know about weather modification, particularly in orographic cloud augmentation. This position is summarized by saying that since we can learn by doing, we should proceed, since the benefits certainly outweigh the risks. The second school claims that the scientific uncertainty is so overwhelming and the well of ignorance so deep that it would be irresponsible to do more than expand controlled research and experimentation, because it is possible that the risks outweigh the benefits.

The "learn-by-doing" school and the "wait-until-all-facts-are-in" school often conflict. Consider, for example, the differences in opinion about use of DDT and other pesticides and toxic chemicals, about flood control and hydroelectric projects like the infamous Aswan Dam, about generation of electricity through nuclear power, or about environmental safeguards necessary during oil shale production. These issues are exemplified by the *Reserve Mining* case[1] so much in the news, where a chief issue is the degree of certainty we must have about the carcinogenic nature of asbestos-like fibers before we stop releasing them into air and water.

It is apparent that the conflict between these schools of thought generates considerable political pressure in governmental agencies, either to accelerate application or to decrease the present rate. An obvious area of scientific, social, and legal uncertainty involves the extra-area effects, since weather modification activities may trigger consequences up to 150 miles (240 km) downwind of the target area. Incidentally, I would appreciate having someone define "downwind," because the definition of this term is exceedingly important. The litigation that could arise is almost beyond the imagination, because it could create justifiable controversies between farmers and urban dwellers, ranchers and skiers, and energy developers and wilderness lovers, to mention only a few. One phrase recurs throughout the literature in reference to extra-area effects: "Robbing Peter to pay Paul."

One reason our knowledge of the societal effects of this technology lags so far behind the technology itself is that it is much more expensive to prove the weather modification hypothesis than it is to apply it. It is often easier to obtain money for operations than it is for research, especially if the state or federal legislatures become involved during periods of concern—such as at present when we have a drought in the Great Plains.

What are some of the legal implications of scientific uncertainty? The popular assumption is that the entire scientific-legal interchange takes place in a courtroom, an idea that should be dispelled. Most lawyers whose practice involves working with aspects of the science and technology related to weather modification never take part in litigation. Their activities afford them an entirely different perspective from the one they would have if they were litigating in a courtroom and doing the necessary procedural paperwork. Some lawyers work in legislative arenas, some are employed as lobbyists on different sides, some are law teachers who can engage in spinning legal theories about the implications of weather modification, and some are agency administrators with responsibility for implementing laws relating to weather modification. Thus "expert testimony" in court is but one situation where lawyers must deal with the issues.

We establish "facts" under legal rules of evidence that perhaps must be adjusted to accommodate scientific standards of certainty or uncertainty. Establishing statistical probabilities does not per se prove an absolute physical "fact." To say there is a 0.7 probability that a certain operation will result in 2 centimeters of additional precipitation also suggests a 0.3 probability that the operation will fail to increase precipitation by that amount. Maybe we should consider simulation modeling as a better tool than probability theory. However, we are going to deal with probability theory in weather modification for some time to come, because we do not have modeling tools available, as Lewis Grant has already pointed out today.

The law does not have to be viewed as a major constraint on weather modification, because lawyers always are faced with the task of pouring the new wines of technological advancement into the old flasks of legal theory. Existing legal theories and institutions accommodated aircraft, electronic communications, computers, spacecraft, and a variety of other innovations.

We also must remember that the law is a shield as well as a sword. The law will not be used only to allege that harm resulted from modification attempts. It can also be used to shield the weather modifier by protecting his expertise and his efforts and the water he gains. The extra-area effects seem to have the greatest potential for social and economic catastrophic effects that could result in legal action. I do not suggest, however, that such possibilities should foreclose attempts to modify weather altogether.

The greater the scientific uncertainty, the greater the difficulty of dealing with the public. This spawns litigation that cannot be squelched by newspaper advertisements, television commercials, interviews on talk shows, and other public relations activities.

The credibility gap is widening, and I hope the weather modifiers do not emulate some representatives of the nuclear power industry by forcing it to become even wider (which brought public initiatives in Colorado, California, and other states with respect to nuclear power). One credibility problem of increasing magnitude concerns possible environmental effects. For example, stabilized precipitation—i.e., fewer floods and less drought—could have both beneficial and adverse effects on terrestrial and aquatic organisms and could decrease both salinity and sedimentation caused by gully-washers in the southwestern United States. Substantial augmentation of precipitation apparently would favor species adapted to water habitats and thus shift the ecological diversity in a particular region. On the other hand, some alpine regions might become even drier because increased snowpack over longer periods of time could mean less available water at critical times.

We cannot assume safely that undesirable effects could be reversed merely by suspending the modification effort. Returning natural ecosystems to former conditions would require a long time, perhaps decades in the southwestern United States. This is not necessarily true of cultivated ecosystems. Short-term ecological studies, like those of the San Juan Ecological Study Project, certainly do not prove the absence of long-term effects. We do not have nearly enough data from these fragmentary studies to establish whether significant long-range effects of persistent weather modification efforts exist. A major controversy brewing in the West concerns modification of orographic clouds in wilderness areas. If this activity were deemed to violate the Wilderness Act of 1964[2] by a court or legislated to do so by Congress, 80% of the weather modification activity in that area might become illegal.

The great challenge we face is learning how to balance scientific uncertainties with political decision making. We have not found a satisfactory way to date. Some persons would like to have a scientific dictatorship and let the scientists make decisions based upon their own data. Others believe that all decisions should be made through some public participation process, that decisions should be based primarily on the number of hands that are raised at a public hearing.

We can no longer afford to make decisions affecting major resources in this country and in the world on the basis of public opinion polls, regardless of which group—environmental, commercial, industrial, or other special interest group—is giving its opinion. At the same time, however, public opinion is a critical component in decision making in a democratic society. How do we balance these factors?

It seems to me that the decision is ultimately going to be political and that both lawyers and scientists should be working toward ways of translating, interpreting, and communicating scientific information to the public, especially to the politicians. There always will be adverse reaction to any difficult decision, but a better informed public would result in a more logical overall process. Otherwise, the decision-making process might deteriorate, and weather modification might be governed solely by public fears that it might cause floods, earthquakes, avalanches, and plagues of locusts.

We must strive to avoid that.

Notes

1. Reserve Mining Co. v. Environmental Protection Agency, 514 F.2d 492 (8th Cir. 1975).

2. 16 U.S.C. secs. 1131-36 (1970).

The Scientific Uncertainties:
A Scientist Responds
Louis J. Batten, Director

Institute of Atmospheric Physics
University of Arizona
Tucson, Arizona 85721

To set this important topic in proper perspective, I begin with a few remarks about why we want to change the weather. There are two general reasons.

First, violent weather kills a great many people and does enormous property damage. A single hurricane that struck East Pakistan in November 1970 killed more than 250,000 people in a single day. Hurricane Camille hit the United States in 1969 and did approximately $1.5 billion worth of damage.[1] An outbreak of tornadoes in the Chicago area on Palm Sunday of 1965 killed about 250 people, and the tornadoes of April 1974 did likewise.[2] Storms kill people and damage property, and it is reasonable to ask whether it is necessary for us to accept this type of geophysical destruction. I say, "No, it is not—it should be possible to do something."

Second, weather modification involves, and in some respects might control, the production of those elements we need to survive. Water and food are currently in short supply in many areas and these shortages almost certainly will be more severe in the future. We can develop new strains of wheat and rye and corn and soybeans and rice, but all is for naught if the weather fails to cooperate. If the monsoons do not deliver on schedule in India, residents of that country starve in large numbers. And if the drought that people have been predicting for the last several years does spread over the Great Plains, there will be starvation around the world on a scale never before experienced.

Weather is the one uncontrollable factor in the whole business of agriculture. Hail, strong winds, and floods are the scourges of agriculture, and we should not have to continue to remain helpless in the face of them. It may be impossible for us to develop the kind of technology we would like to have for modification of weather, but to assume failure in such an important endeavor is a course not to be followed by wise men.

The most significant fact that Lewis Grant mentioned in his presentation is that we can change the weather. As he pointed out, a cloud is composed of billions of tiny water droplets, the biggest of which is about the diameter of a human hair but most are perhaps only a tenth of that size. We cannot see the individual cloud droplets but we can see the aggregate of billions of them. Because the water is so pure, the droplets often do not freeze when their temperatures are less than $0°C$. They then are called supercooled. If nature or technology introduces ice crystals into these clouds, the ice crystals will grow according to the law of physics, and the aggregation of ice particles will form snowflakes that fall out of the cloud. If it is warm near the ground, the snow melts and we witness rain. Most of the precipitation that falls to earth follows this pattern.

There are many clouds composed of these tiny, supercooled water droplets that might have a temperature of $-10°C$. Owing to their nature, they do not produce precipitation of any kind. They are born, they last for minutes or several hours, and they evaporate. It was discovered about 30 years ago that the addition of dry ice—because it is very cold—or a finely divided chemical such as silver iodide produces ice crystals that will initiate the growth of snowflakes and raindrops.

That experiment has been replicated many times around the world, and there is no question about what happens. The question is whether we can produce enough additional rain or snow to be of any consequence. A few snowflakes are not of much value. Can we increase rain over farmlands sufficiently to influence agricultural yields? Can we significantly increase water supplies for hydroelectric power production?

This issue has been addressed over the years, and the answer one receives depends to a certain extent on the person asked. It also depends on how we define proof. A scientist often says, "I want to see proof that comes from sound scientific procedures and experimental design and data analysis." Scientists use words like "randomized experiments" and "statistical significance" and set criteria for proof of efficacy of cloud seeding. Evidence of our ability to increase rain by 20% is scarce, although some exists. Many experiments encourage reasonable persons to conclude that additional snow had been produced or that more rain fell, but the quantitative evidence is limited.

On the other hand, many believe that cloud seeding works, which obviously is different from proof. A recent federal report summarizes weather modification experiments in the United States and states that at least 45 operations for increasing rain or snow or for reducing the amount of hail were conducted in fiscal 1973 over about 135,000 square miles (349,650 km^2).[3] So, during that year something on the order of 5% of the land area of the United States was involved in cloud-seeding operations for some purpose.

Industrialists, government leaders, and scientists in about 30 countries now involved in this kind of activity are convinced that investment of funds and time in cloud-seeding programs is prudent. Many people believe that it works and that is an important point to ponder.

A key question is, "Did investors receive adequate returns on their investments?" We now move from the factual area to that of speculation. In many cases no one will ever know because of the difficulty of separating the activity of the modifier from the activity that would occur naturally. If nature is good to us during the cloud-seeding operation and provides a wet period, there is a tendency to believe that the money was well spent. On the other hand, if drought happens to prevail during the experimental period, the investors tend to be upset. The difficulty with many commercial operations—but by no means all—lies in arriving at convincing conclusions as to their efficacy. I hasten to point out that data from a number of carefully done commercial seedings strongly suggest that the person who paid for the operation got a fair return on the investment. In many other operations and experiments, it is impossible to tell. It really is somewhat like going to a physician when you are not feeling well. You receive an examination and a prescription and, if three days later you feel better, you figure you got your money's worth.

There are a few other important facts in this business. We must recognize that clouds and storm systems last from minutes to close to a week. Small cumulus clouds might last only several minutes and then evaporate, summer thunderstorms might last 30 to 60 minutes, and hurricanes might endure for the better part of a week. In interpreting the results from experimental weather modifications, one must keep this in mind. After a storm system has been seeded and precipitation has occurred, it is reasonable to inquire about what happens to the potential precipitation in that storm after it passes the target area. This is difficult to answer conclusively.

This leads to the important question of who owns the rain. Every drop of water on the ground or under the ground in the western United States belongs to somebody. But who owns the precipitation still in the atmosphere is a question of great importance because it can mean the difference between success or failure of a crop or other enterprise. It is particularly important because the most beneficial rain during the growing season falls mostly from thunderstorms that last a rather short period of time and that do not provide uniform rainfall over a large area. In an article a few years ago, entitled "It Rained Everywhere but Here—the Thunderstorm Encirclement Illusion,"[4] the author observed that a farmer in an area over which there are widely scattered thunderstorms believes that many other farmers are receiving rain. But thunderstorms may cover only a very small fraction of the area. These large storms extend up to 13,000 or 16,000 meters, so thunderstorms may be seen in many directions, yet only limited areas receive the rain that is falling.

This leads to obvious problems for advocates of a seeding operation. One landowner may receive 2 centimeters of rain while a neighbor receives 2 hours of sunshine. It also raises the question of whether seeding a cloud in one locale causes increased precipitation downwind—on the next farm, on the next county, on the next state, or even on the next country.

This raises the final topic I would like to mention—the international problems associated with weather modification. Many persons have been actively concerned with this problem over the years, and it is more complicated than whether a cloud-seeding program over the Great Lakes might reduce rainfall in Canada—a legitimate question but a relatively simple one on the broad scale of things. The important issue now facing us is our national effort to control hurricanes.

Hurricanes are exceedingly important for many reasons. They do much damage and much good. Newspaper headlines tell about the destruction and not about the beneficial influences. The whole ecology of the southeastern United States would be different from what it is today if hurricanes suddenly stopped coming over that part of the country. Also, Southern California and Arizona survive as well as they do because of rain from the periodic hurricanes that form to the west of Mexico and move to the southwestern states. We would like to have a technique to lower the wind velocity and thus to reduce the damage done by the physical force of the wind and by the ocean water blown over the coastline, while at the same time not reducing total rainfall.

This is hard to do but it is a reasonable goal, and some empirical evidence suggests that it is possible. An experiment called Project Stormfury[5] included plans to seed hurricanes in the western Pacific in 1977. Unfortunately, the governments of Japan and mainland China do not want us to perform the experiments. As a result, the tests have been delayed until a suitable site is found. It is anticipated that international legal or political problems will become even more evident in the future.

A related aspect is use of weather modification activities during wartime. Project Popeye in southeast Asia involved cloud seeding in an attempt to increase rainfall over the Ho Chi Minh Trail as a way to reduce infiltration of troops and equipment.[6] The United Nations now is considering a resolution that dedicates weather modification activities to peaceful purposes.

So much for the problems concerning weather modification. We must learn to modify the weather beneficially. Evidence to date suggests that we can, but it will require a major research effort. The research program in the United States has been retrogressing for the last few years and, indeed, is almost designed for failure. Unless the trend is reversed, it assuredly will be a failure, and people all over the world will pay the penalty in terms of property losses and starvation.

JAMES CURLIN: I gather that we are still at a rudimentary level of knowledge about modification techniques. Is there any futuristic higher technology? Are there other methods of balancing the energy equations or of altering the energy relationships other than by cloud seeding? How do we reduce wind without influencing precipitation?

LOUIS BATTAN: Most of the efforts have concentrated on seeding because it is possible to construct a physical model for expectations of what will occur. Also, experimental evidence suggests that in the past we have had some success in augmenting precipitation. We have seen since about 1960 a series of reports on hail suppression in the Soviet Union that announce spectacular success.[7] They use artillery guns to introduce ice crystals into cloud systems, with the assistance of radar to locate thunderstorms. They also use rockets of about 15 centimeters in diameter to fire materials into clouds. This technology has not been used in the United States for obvious reasons, or at least obvious to a person flying a small plane over Oklahoma. But this is how technology progresses, by different people attempting to solve problems in different ways.

It is in this time-tested manner, perhaps, that we will be able to distinguish effects on wind and on precipitation. The National Academy of Sciences reported recently that it might require about 10 years of intensive research to develop an adequate technology for precipitation augmentation.[8] Some feel that this is unrealistic. My impression is that it will require about that long to design the proper experiments, to analyze the data properly, and to distribute the results effectively.

I hope we will have sufficient opportunities during the discussions today and tomorrow to consider these matters further.

Notes

1. National Oceanic and Atmospheric Administration. 1969. Hurricane Camille. Report to the Administrator of the National Oceanic and Atmospheric Administration.

2. National Oceanic and Atmospheric Administration. 1974. The widespread tornado outbreak of April 3-4, 1974. Natural Disaster Rep. 74-1.

3. Domestic Council. 1975. The Federal Role in Weather Modification. Washington, D.C. 33p.

4. McDonald, J. E., 1959. It rained everywhere but here—the thunderstorm encirclement illusion. Weatherwise 12:158-60.

5. *See* Annual Reports, Project Stormfury. National Hurricane and Experimental Meteorology Laboratory, N.O.A.A., Miami, Fla.

6. Shapley, D., 1974. Weather warfare: Pentagon concedes seven-year Vietnam effort. Science 184: 1059-1961.

7. Battan, L. J., 1973. Survey of weather modification in the Soviet Union: 1973. Bull. Am. Meterol. Soc. 54: 1019-30.

8. Committee on Atmospheric Sciences, National Research Council. 1973. Weather & Climate Modification: Problems and Progress. National Academy of Sciences, Washington, D.C. 258 p.

Legal Uncertainties of Weather Modification*

Ray J. Davis

School of Law
University of Arizona
Tucson, Arizona 95721

> The law is a sort of hocus-pocus science, that smiles in yer face
> while it picks yer pocket; and the glorious uncertainty of it is of
> mair use to the professors than the justice of it.[1]

Introduction

Legal uncertainty can be intentional. A Roman emperor had laws carved at
the tops of high columns, and many law professors are alleged to enjoy
vagueness in Socratic dialogue that never reveals answers. Doubt also can arise
by the unavailability of necessary legal materials. The Supreme Court argument in the "Hot Oil" case[2] produced admission by the government attorney
that the only copy of the code of fair conduct in question had been seen in
the hip pocket of someone in Texas.[3]

Weather modification law can be found—it is neither hidden nor unavail-
able. Rather, the problem stems from two other factors: (a) infrequency of
litigation leaves us without adequate judicial clarification of the law, and (b)
scientific uncertainties make the facts to which the law is to be applied not
fully predictable.

Factual uncertainty may cloud human perception of reality. It has been
noted:
> Appearances to the mind are of four kinds.
> Things either are what they appear to be;
> or they neither are, nor appear to be;
> or they are, and do not appear to be;
> or they are not, and yet appear to be.
> Rightly to aim in all these cases
> is the wise man's task.[4]

The hazard of factual misperception can be reduced by discussing the law in the context of scenarios based on weather modification situations. Four scenarios will raise the relevant legal issues:

1. Precipitation enhancement,
2. Hail suppression,
3. Runoff augmentation, and
4. Tropical storm modification.

The discussion of each scenario seeks out the potentially applicable legal norms and their possible interpretations. Attention is directed also to other issues of legal uncertainty. What other norms could be applied to the supposed facts? What should be the law? How should the diverse interests of the general public, the consumer of weather modification, the seeder, the others affected, and the professors whose business is the glorious uncertainty be accommodated?

Precipitation Enhancement

Scenario

To someone coming from Arizona, as I do, the need for more precipitation in southwestern Minnesota seems quite minimal. Yet to some local inhabitants of that area, rainmaking from silver iodide seeding of summer cumulus clouds appears to be a necessary and practical way to increase farm productivity.[5] In 1969, nine named counties were given permission by the Minnesota legislature in a special law to enter into contracts of up to $5,000 yearly with weather modification firms.[6] According to local procedure, it requires adoption and filing of a county resolution for such legislation to become effective.[7] Five counties did file to bring themselves within the law.[8]

Such limited operational authority, coupled with the complete lack of any regulatory provisions in the Minnesota law, is of concern to those persons in the state who wish to increase the possibility of using artificial precipitation enhancement as an agricultural tool while at the same time protecting user groups and the public from being victimized by incompetent or unscrupulous rainmakers. The Minnesota Department of Agriculture formed a special Agriculture Weather Modification Task Force, which held hearings throughout 1975, both in southwest Minnesota and in St. Paul. Several versions of control legislation were drafted, and the legislature considered whether the state should have a law authorizing government-sponsored cloud seeding on a wider basis and creating a regulatory system to control governmental and commercial rainmaking.[9] No weather control law was enacted by the 1976 Minnesota legislature.

Under present Minnesota law, anyone, regardless of cloud-seeding skills, can embark upon a weather modification effort. And such an operation can be continued unless the seeder runs afoul of aeronautic or environmental rules[10] or causes harm.[11] The task, then, before future Minnesota legislatures is to pass a bill that will reduce the legal uncertainties of weather modification in the state and, should the need appear to be great enough, provide some broader system for public financing of precipitation enhancement.

Regulation

Minnesota and other jurisdictions seeking a legal basis for regulation of weather modification have available a wide array of existing state laws from which they can select particular norms. An analysis of state laws in effect on August 1, 1975, describes several of the components of a weather modification legal regime[12] as follows.

1. State agency to administer weather modification legislation

This component refers to the state agency with ultimate authority and responsibility to administer state weather modification control legislation. In 14 states the umbrella agency was concerned primarily with natural resources or environmental matters. This category includes water commissions,[13] natural resources departments,[14] an ecology department,[15] and an office of environmental affairs.[16] Four state agriculture departments administer weather modification legislation.[17] The present interest shown by the Minnesota Department of Agriculture suggests that it would be the logical agency for that state. Six states had other governmental entities handling such matters.[18]

2. Weather modification board to be created

Legislation in 13 states creates special weather modification boards of one of three types: (a) independent weather modification boards;[19] (b) weather modification advisory boards to give advice to the state agency administering weather control legislation;[20] and (c) weather modification boards with policy and decision-making powers which report actions to the state umbrella agency.[21]

Independent boards are out of fashion. They involve too much overhead and too small a workload to justify themselves even in prosperous times. The two practical choices for Minnesota are an advisory board or a board with actual decision-making powers.

3. Professionals to be regulated (licensing)

This concerns issuance of state certification to individuals for the privilege of practicing weather modification. It does not relate to provisions under which the power to conduct precipitation enhancement operations in a specified area for a specified time period is granted. Thirteen states now require seeders to obtain professional licenses before practicing their trade.

Two comments should be made here: Criteria for issuance of licenses vary widely among the states;[22] and such professional licensing has been sought by weather modifiers rather than thrust upon them.[23] The person with established competency and a solid reputation has the most to gain by barring incompetent operators.

4. Projects to be regulated (permit issuance)

This element of weather modification law involves granting franchises for particular projects. Twenty-three states require either registration of projects or operational permits.[24]

The permit feature in a state law is of critical importance in regulating precipitation enhancement. Properly drafted legislation or administrative rules will require enough information from applicants to allow agency staff to judge whether the project is soundly conceived.[25] Appropriate conditions and limitations about timing and methodology can be incorporated into the project permit.[26] Modification, suspension, and revocation of permits can reinforce the notion of governmental supervision of the precipitation enhancement activity.[27] Of course, even if Minnesota should enact adequate safeguards, a great deal will depend upon the ability of the agency staff to enforce them adequately.

5. Prior public notice to be given

Of states with a permit requirement, 18 require persons intending to undertake weather modification operations to provide public notice of that intent before being granted a permit.[28] In one jurisdiction, the regulatory agency is empowered to determine whether such notice shall be required.[29] Unfortunately, notice provisions by themselves are more apparent than actual ways of informing persons affected by a proposed precipitation enhancement operation.

6. Public hearing to be held

Five states require regulatory agencies to hold public hearings in the area to be affected by an operation prior to granting an operational permit.[30] In three other states, the regulatory agency is authorized to do so,[31] with the rationale that some permits involve such minor projects that the *de minimus* principle applies—that is, it just is not worth the effort to solicit public reaction. It should be recognized that even in states with mandatory hearings, popular opinion can be ignored by the agency; the views expressed at a hearing may or may not really be *vox populi*.

7. Records to be maintained

For effective regulation, the responsible government agency needs information about the activity regulated. Twenty states now require operators to keep weather modification records and/or to report operational information to the regulatory agency. To some extent such requirements overlap the federal statute[32] and regulations of the National Oceanic and Atmospheric Administration requiring cloud seeders to report.[33] Minnesota could follow the example of some states and insist merely on receiving copies of the federal reports.[34]

Operations

Weather modification research and development is funded primarily by the federal government.[35] Operational programs in North Dakota,[36] South Dakota[37] (both neighbors to the west and upwind of Minnesota), and Utah[38] have been mostly state supported. This governmental support of cloud seeding must be based upon legislation. Congress and the state legislatures exercise the power of the purse both by collecting money under their taxing powers and by dictating how to expend it.[39] Two types of legislation are involved in expending government funds:

1. Authorization statutes

Nineteen states have special provisions that authorize use of public funds for weather modification. These funding laws are of three types: (a) The agency that administers weather modification legislation is authorized to spend public funds for conducting or sponsoring weather modification activities;[40] (b) existing state or local agencies are authorized to spend funds for conducting or sponsoring these activities;[41] or (c) special weather modification districts are created with power to tax and spend.[42]

North Dakota and South Dakota give weather modification agencies the authority to contract for cloud seeding in areas where local entities elect to participate during a particular year. Thus, they have a combination of types (a) and (c).[43] Minnesota might opt for that model, or it might select the route followed by most of the High Plains states—legislation authorizing local weather modification districts.[44]

However, it probably is more feasible politically for Minnesota to adopt a type (b) statute—as California did—and authorize existing water resources development agencies (*e.g.,* cities, counties, special districts) to use their funds for weather modification. This in essence is what the present Minnesota law does, except that it restricts the authority to only a few counties and places an unrealistic limit on the amount that might be expended.[45]

2. Appropriation statutes

Authorization alone does not provide the power to spend what has not been appropriated. The power of legislative bodies to encourage or inhibit precipitation enhancement is manifested most dramatically by the amount of money appropriated expressly for that purpose or appropriated without restrictions that would bar it from being spent for that purpose. The Senate in South Dakota in 1976 eliminated the statewide seeding program by failing to pass the appropriate bill necessary for its continuation.[46]

Hail Suppression

Scenario

During the early 1950s, weather modification enthusiasts sometimes oversold their capacity to alter storms that produced damaging hail. They spoke of "cloud-busting" and left a lasting impression that successful hail suppression reduces precipitation. This, coupled with the notion that weather modification involves "robbing Peter to pay Paul," has produced in some parts of the nation an abiding antagonism toward weather modification in general and hail suppression in particular. Nowhere has the dislike of the "hail-chaser" been more profound than in the tri-state area of Pennsylvania, West Virginia, and Maryland. Although there is some support for the view that seeding of supercell storms *sometimes* reduces precipitation (negative precipitation),[47] members of the Tri-State Natural Weather Association believed that clandestine and illegal hail suppression efforts had worsened drought conditions.[48]

Legislative activity in the three states reflected public sentiment. A Pennsylvania township passed an ordinance banning cloud seeding that was validly applied to the operator of a ground-based cloud-seeding generator.[49] Maryland legislatively barred all artificial cloud modification on a temporary basis, extended the statutory prohibition, allowed it to lapse, and then, for good measure, repealed it[50]—but only after seeding in the state had stopped. The current law in Pennsylvania,[51] which West Virginia adopted in essence,[52] constitutes a *de facto* ban on weather modification activities. Its provisions clearly oppose cloud seeding, and it has succeeded, unless secret seeders are at work.

Of the 15 major lawsuits filed concerning weather modification,[53] 7 dealt with hail suppression,[54] and 2 of those arose in the tri-state area.[55] These hail suppression cases further demonstrate the legal uncertainties of weather modification. In litigation over legal liability or injunctive relief, it is necessary under our system to prove that a legal norm has been violated. Complainants have been none too successful, prevailing in only one criminal case from Pennsylvania[56] and one civil case in Texas.[57]

A tongue-in-cheek comparison of the common law system of the English-speaking world with the legal systems of Germany, Russia, and France declares:

> The common law regards everything as legal,
> Unless statute or case law makes it illegal;
> Under German law everything is illegal,
> Unless the law specifically makes it legal;
> According to Russian law everything is illegal,
> Even though the law says it is legal;
> But in France everything is legal,
> Even though the law says it is illegal.[58]

Although this comparison unduly disparages the other three legal systems, it is a workable starting point for comprehension of the Anglo-American common-law system. Conduct in the United States is lawful and cannot be penalized or enjoined unless some statute, administrative action, or court decision declares otherwise. Weather modification activities are lawful if not made unlawful.

A survey of the lawsuits concerning hail suppression indicates that to recover damages or to obtain an injunction, some person or group from the tri-state area would have to prove:

1. Harm to a legally protected interest,
2. Conduct by the defendant that can be characterized as faulty or can be classified as involving an activity for which there is liability without regard to fault, and
3. A causal relationship between the conduct of the defendant and the harm to the plaintiff.

Moreover, even if these are satisfied, two other rules apply:

1. The defendant will prevail if able to establish a defense, and
2. There are legal restrictions on who can sue and who can be sued.

These five points are discussed below.

1. Harm to a legally protected interest

To recover for harm, the plaintiff must allege and prove specific injuries that are compensable. Courts consistently deny recovery for emotional strain, anxiety, and fear.[59] They have, however, entered judgments for persons who have suffered personal or property injury.[60] Consequently, if property loss occurs as a result of a suppression effort, the harm is the sort that can be compensated. If precipitation is decreased through seeding for hail suppression—the complaint of the people in Pennsylvania, West Virginia, and Maryland—compensation or injunctive relief can be granted if, but only if, the plaintiff is regarded as having a property right in natural precipitation. On that issue the few cases conflict. One from Texas in the 1950s says "Yes,"[61] an earlier one from New York asserts "No,"[62] and in the late 1960s a lower court in Pennsylvania agreed that landowners beneath clouds had a property interest in the precipitation from them but decided that this interest would give way to properly approved seeding in the interest of the general public.[63]

Three states have statutory norms concerning atmospheric property rights. Colorado says, in effect, that all moisture suspended in the atmosphere is the property of the people of the state and is subject—like surface and underground waters—to being used under a right gained through prior appropriation.[64] The Utah statute reads:

All water derived as a result of cloud seeding shall be considered as part of Utah's basic water supply the same as all natural precipitation water supplies have been heretofore, and all statutory provisions that apply to water from natural precipitation shall also apply to water derived from cloud seeding.[65]

That provision can be read as denying any special right to claim atmospheric moisture. North Dakota, the third state, declares:

[W]ater derived as a result of weather modification operations shall be considered a part of North Dakota's basic water supply and all statutes, rules, and regulations applying to natural precipitation shall also apply to precipitation resulting from cloud seeding.[66]

None of these three provisions has been tested judicially. None of them deals specifically with the right of a property owner who asserts harm to his interests as the consequence of hail suppression. Inferences can be drawn that perhaps plaintiffs in these three states would have a difficult time asserting that a rain shadow allegedly caused by hail suppression deprived them of some property interest. Nonetheless, the law's lack of certainty remains.

2. Liability of defendant

The second element of the plaintiff's case is proof either of fault or of conduct that imposes liability without fault. Fault in the hail suppression context could be established by proving carelessness or malpractice by the seeder. A professional is under a duty to perform with such care, skill, and diligence as persons in that profession ordinarily exercise. Failure to do so is negligence.[67] To determine if a professional did not satisfy the standards normally met by members of that profession, expert witnesses should be called to answer questions regarding scientific knowledge.[68] Plaintiffs have encountered difficulty in weather modification litigation due to an inability to muster sufficiently impressive expert witnesses.[69] Obviously, that could change in the future. The professionals usually have rallied around the defendants, but that, too, might change.

In two of the states, West Virginia[70] and Pennsylvania,[71] it is not necessary to prove fault. Claimants merely must prove that their losses have been caused by droughts or floods induced by cloud seeding. In Maryland and some other states, plaintiffs need not prove fault if they can prove that hail suppression is abnormally dangerous or ultra-hazardous. The basic element of this determination is whether the risk created by the activity is great and cannot be eliminated by the exercise of reasonable care.[72] Here courts might distinguish between seeding supercell storms and other storms, with the argument that scientific evidence shows that seeders cannot eliminate the great risk of loss caused by modified supercells. Hence, the activity is abnormally dangerous and the modifier is liable even if not negligent.

It should be noted that statutory provisions in a few states declare that weather modification does not constitute an ultra-hazardous activity.[73]

An alternative method of evading the fault requirement is to establish that the suppression constituted a private nuisance. The law of nuisance balances the harms likely to be caused against the benefits likely to result and then determines whether on balance the activity unreasonably interfered with the plaintiff's right to the use and enjoyment of property.[74] No court has struck the balance for hail suppression. In one case involving precipitation enhancement, the court's language tends to support the proposition that development of atmospheric water resources on behalf of a municipality does not constitute a nuisance to property owners who might suffer from additional precipitation.[75] That, of course, leaves considerable doubt about what a court in another jurisdiction would do after balancing an alleged decrease in precipitation against the diminution of damaging hail, especially if only private parties were involved.

3. Cause and effect

Failure to demonstrate the linkage between conduct of the defendant and harm to the plaintiff's property remains the major impediment in litigation involving hail suppression and other types of cloud seeding. In fact, the basic reason why plaintiffs lost the civil case in Pennsylvania was their failure to establish causation.[76]

The plaintiffs in only one case succeeded in establishing the required causal link. The case arose in Texas in the 1950s and involved conflicting expert testimony. The court relied upon lay testimony to establish the necessary connection between loss by the plaintiffs and seeding by the defendants.[77] As evidence accumulates on efficacy of hail suppression, the chances increase that plaintiffs will be able to prove causation. In a 1974 Texas case, however, they did not succeed because the defendants' experts overwhelmed them. The defendants were helped also by having a valid operational permit under legislation[78] passed after the earlier Texas litigation.

It should be stressed that adequate proof of causation requires more than establishing that a particular activity occurred at a given time or that hail suppression really is an effective technique. It must be established that a particular series of events was a necessary antecedent to the harm and that without the seeding the loss would not have occurred.

4. Defense against the charge

Because the plaintiffs have failed to establish their cases, the defendants have not needed to advance and prove valid defenses to escape liability. As noted above, this may change, and defendants will be pressed to establish defenses. In limited situations, they could demonstrate that the plaintiff was contributorily negligent[79] or consented to the risk involved,[80] but a more likely defense would be that of privilege. A person has a legal privilege to protect his property by certain appropriate means in particular situations. An example is the right of a farmer in some jurisdictions to fend off surface waters without being required to take into account consequences to others, who also can protect themselves as best they can.[81] Perhaps there is an analogy between this "common enemy" doctrine and a privilege to use weather modification to stop hail damage to lands and crops.

Another defense is sovereign immunity. Although many states have abolished the doctrine, some states and their agencies remain immune from liability for losses they negligently cause.[82] The federal government, through the Federal Tort Claims Act, partially waived its sovereign immunity. An important exception to the waiver is any so-called discretionary function of the federal government.[83] Case law holds that weather forecasting is within the exception, because it involves daily decisions that include policy and judgment as well as discretion in predicting and disseminating reports about the weather.[84]

It might be argued by analogy to the forecasting cases that there will be no federal liability under the Federal Tort Claims Act for losses caused by faulty planning of hail suppression projects.[85] It is possible, however, that losses due to operational miscues will not be within the discretionary function exception, and that federal liability will result.

5. Parties to a suit

Litigation is so expensive that the average landowner cannot afford to sue a weather modifier. It is important that the plaintiff be joined by other injured persons so that they can pool their financial resources. A common procedure for this is the class action suit.[86] The concept of using a class action to stop or to collect damages from a hail suppression project is feasible if the project is of a large scale. But many technical requirements must be met by the plaintiffs to institute such an action. Since World War II, the trend generally has been toward reducing the difficulties of bringing class actions.[87] In the past few years, however, federal courts have reversed the trend and have rendered decisions that decrease the vitality of the class action concept.[88] The entire issue of hail suppression, including legal and institutional arrangements, is given comprehensive treatment in a forthcoming technology assessment.[89]

Runoff Augmentation

Scenario

The economic vitality of the southwestern United States depends on availability of water from the Colorado River system. But should Mexico and each of the upper and lower basin states insist on using what has been awarded to them by treaty, compact, statute, and case rulings, there will be insufficient water to meet the entitlements. So the question arises about ways of increasing the flow of the Colorado. One technique is increased snowpack in the Rocky Mountains to augment runoff. Assuming that calculations from Bureau of Reclamation experiments can be witnessed in an operational project, the technological prospects are promising indeed.[90] But some stubborn legal problems will not go away, among them interstate regulation of weather modification.

Interstate Problems

The means of regulating weather modification and allocating streamflow on an interstate scale can be grouped in six major classes.

1. Uniform or reciprocal legislation

Uniform legislation, whereby states within the Colorado River Basin enact similar laws with respect to allocation of supplemental water derived from snowpack enhancement, would promote uniformity. Such legislation could be proposed to the state legislatures by a body such as the Council of State Governments,[91] or perhaps model legislation could be drafted by the National Conference of Commissioners on Uniform State Laws.[92]

An alternative method is reciprocal legislation, in which states take into account the law of their neighbors. The Utah Cloud Seeding Act contains such a provision. It requires dual compliance by stating:

> Cloud seeding in Utah to target an area in an adjoining state is prohibited except upon full compliance of the laws of the target area state the same as if the cloud seeding operation took place in the target area state, as well as the other provisions of this act.[93]

A kind of negative reciprocal legislation could govern operations intended to affect flow in a downstream state. An example is a Colorado provision:

> Operations cannot be conducted in Colorado for the purpose of affecting weather in any other state if that state prohibits such operations to be carried on in that state for the benefit of Colorado or its inhabitants.[94]

New Mexico, another Colorado River Basin state, has a similar "least favored nation" clause in its cloud-seeding control law.[95]

2. Equitable apportionment

The doctrine of equitable apportionment of water may be applied only in suits between states,[96] and the rights of individual water users will not be determined by the litigation.[97] The eleventh amendment will not bar a suit by one state against another where the controversy is more than "a mere question of local private right and involves a matter of state interest."[98]

Equitable apportionment is used by the Supreme Court of the United States to divide the water in issue among litigating states by balancing all relevant factors to reach a fair and equitable determination.[99] The doctrine could be applied to resolve interstate conflicts over allocation of moisture in the atmosphere or of supplemental water in interstate streams—i.e., water allegedly added by melt from artificially augmented snowpack. Any such application of the doctrine would affect considerably the interstate regulation of operations, as well as allocation of the water.

3. Congressional allocation

In *Arizona v. California*,[100] the United States Supreme Court interpreted the meaning and the scope of the 1928 Boulder Canyon Project Act to determine how much water each basin state could use from the Colorado River and its tributaries. The act authorized the Secretary of the Interior to allocate water among the basin states without regard to state laws and under very broad federal standards. The Court referred to the "general welfare" power in upholding congressional legislation that apportioned water in an interstate stream and its tributaries.

As one writer observed, this congressional power makes possible a fully integrated development-allocation scheme without interference by conflicting state authority.[101] Cloud seeding in the Rocky Mountains could be coordinated with the downstream needs of users in the arid Southwest.

4. Federal management

When issues are of a regional or interstate nature, the states have difficulty solving them either by themselves or together. If they fail to reach agreement, the federal government may act.[102] The problems and developments that gave rise to *Arizona v. California* illustrate how runoff augmentation also could be handled. One mechanism for federal control over streamflow could be creation of a federal corporation,[103] but there seems to be no federal disposition to do this.[104]

Congressional legislation must be based on the delegated powers of the Constitution. The commerce clause, for example, which gives Congress power to regulate commerce,[105] includes control of navigation.[106] This clause encompasses groundwater flowing into navigable waters of the United States[107] and would be adequate authority to regulate atmospheric water resources.[108]

5. Interstate agreements

It is difficult to generalize about the legal meaning of interstate agreements. They are informal and often are designed for the convenience of the parties without binding obligations. One writer observed that they usually are based on personal relationships and can be useful in resolving minor irritations. The device helps avoid red tape and has been used to create regional planning agencies between neighboring states.[109]

Agreements between states do not require the consent of Congress if they do not affect the just supremacy of the United States or its political integrity and do not interfere with its sovereign treaty-making powers.[110] Thus, enforcement of liability through an interstate reciprocity agreement was held valid as not violating the Constitution.[111]

An example of an interstate agreement is the North American Interstate Weather Modification Council (NAIWMC), whose main purpose is "to achieve and maintain state and local control of weather modification activities while endeavoring to attain a high degree of legislative uniformity and an effective information exchange mechanism." States or provinces can be permanent or temporary regular members or affiliates.[112]

Another example is the Contract for Cooperative Weather Modification Research in the High Plains. Under the Memorandum of Understanding, signed by Kansas, Nebraska, Colorado, and the Bureau of Reclamation, the parties agreed to work together toward a program plan with the bureau's assistance. The memorandum can be amended if all parties agree, and any party can withdraw by giving all other parties 120 days' notice of its intent to do so.[113]

Courts might rule that such informal agreements are mere understandings drafted to avoid any commitments harmful to a state's interests. Even if enforced—the remedy could well be arbitration—and whatever the result, the lawsuits would discourage subsequent interstate agreements.

Interstate operations can also be conducted without an interstate agreement. In 1975 a number of counties in Utah and Idaho created the Utah-Idaho Weather Modification Corporation (UIWMC), a nonprofit corporation chartered in Utah. The corporation is supported by the state of Utah, as well as the Utah counties involved. In 1977 the contribution of Idaho's counties will come from newly authorized weather modification taxing districts. Each county in the UIWMC has a weather modification board that determines when cloud seeding should be undertaken.[114]

6. Interstate compacts

States often use interstate compacts in dealing with natural resources.[115] Normally, there are four stages to making a compact: negotiation and drafting, state ratification, congressional consent, and operation.[116]

The binding effect of interstate compacts is illustrated by the leading case of *West Virginia ex rel. Dyer v. Simms*,[117] holding that the Supreme Court could interpret state law bearing on compacts between states and that a state's constitution could not block adherence to the compact. Terms of a compact cannot be amended or modified except by mutual agreement of all party states,[118] and a compact is binding on the citizens of each state and on the judicial and executive branches of the state governments.[119]

In *Application of Waterfront Commission of New York Harbor*,[120] the court upheld the commission's authority to issue subpoenas throughout both states as provided by the New Jersey—New York Compact.

It is clear that compacts truly bind states that are parties to them, and they can be drafted to accomplish the desired purposes. To be effective,

> an interstate compact must create a commission charged with well-defined responsibility, must give the commission adequate authority to establish rules and regulations and to enforce them through legal proceedings, must avoid restriction of action by the commission through voting requirements or the establishment of veto authority by any signatory state, must provide for proper staffing and, finally, must provide for adequate financing.[121]

Aside from interstate compacts there are federal-interstate compacts, which the National Water Commission in *Water Policies for the Future* recommends "as the preferred institutional arrangement for water resources planning and management in multistate regions."[122] An example of such a compact is the Delaware River Basin Commission (DRBC),[123] created after several years of litigation among the basin states. The DRBC is unique in two respects: (a) The United States is a signatory party with the states, and (b) extremely broad powers are granted to the compact commission. The commission was charged with formulating a comprehensive plan to which all federal, state, local, and private water-project planners must conform. The DRBC has licensing authority and broad regulatory powers as well.[124]

The compact device is versatile, but its proper use would require firm commitments by the party states to interstate weather modification programs. Although antagonisms among the states in the Colorado Basin are less ardent than in the past, achieving a binding legal compact among them relating to augmented streamflow from weather modification will be most difficult. Indeed, the very effort to effect such an agreement might reopen old wounds.

Environmental Concerns

Many persons and organizations, both in and out of government, share serious concerns about the environmental effects of weather modification.[125] Legal responses to these concerns include full disclosure of the potential environmental impact, reporting of plans and operations, and outright bans on augmentation efforts in specified areas.

1. Disclosure

Snowpack augmentation in the Colorado River Basin thus far receives its widest public exposure through the Bureau of Reclamation's Skywater Program.[126] The National Environmental Policy Act (NEPA) mandates federal agencies, such as the bureau, to file environmental impact statements with the Council on Environmental Quality whenever they propose any "major Federal actions significantly affecting the quality of the human environment." The law requires these documents to consist of a "detailed statement" on (a) the environmental impact of the proposed action, (b) any adverse environmental effects that cannot be avoided should the proposal be implemented, (c) alternatives to the proposed action, (d) the relationship between local short-term uses of the environment and the maintenance and enhancement of long-term productivity, and (e) any irreversible and irretrievable commitments of resources involved in the proposed action should it be implemented.[127] This advance disclosure provides the public with information on the potential environmental consequences of snowpack augmentation before the operation begins. The law requires federal agencies to consider environmental matters from the outset and to use this information in making subsequent decisions.

Federal agencies sometimes decide that an activity is not covered by NEPA and do not file impact statements, while in other cases statements have been either substantively inadequate or procedurally not in compliance with the law or regulations promulgated under it. This has resulted in many federal cases that interpret NEPA.[128] Valid uncertainties can arise about whether particular snowpack augmentation programs constitute "major Federal actions significantly affecting the quality of the human environment."

The first step in meeting the NEPA requirements for something as significant as increasing the flow of the Colorado River by the amounts foreseen by the Bureau of Reclamation would be to draft a statement covering all the items listed above. The information for the draft may come from environmental literature, studies prepared for the particular project, and other environmental reviews and assessments. The draft impact statement would be circulated among interested state and federal agencies and concerned groups and persons. Public hearings could be held, and indeed should be for a project as controversial as the proposed augmentation operation. Information derived from hearings and from comments on the draft should be incorporated into a final statement filed with the Council on Environmental Quality.[129]

Some states have followed the lead of Congress by enacting similar legislation requiring the preparation of environmental impact statements.[130] New Mexico, one of the Colorado River Basin states, enacted a "mini-NEPA"[131] but repealed it after further consideration.[132]

2. Subsequent reporting

The present federal statute on reporting authorizes the Secretary of Commerce to regulate reporting requirements for weather modification activities, including cloud-seeding efforts to deepen snowpack.[133] This administrative authority was in turn delegated to the National Oceanic and Atmospheric Administration (NOAA), which published regulations in the *Federal Register* in 1972.[134] Subsequently, NOAA officials, acting under a White House mandate to require more environmental information, drafted proposed changes. After discussions with representatives of the weather modification industry, more limited rules were promulgated and went into effect in 1974. They request cloud seeders to disclose whether they have filed impact statements and, if so, to supply copies to NOAA. Other questions are: "Have provisions been made to acquire the latest forecasts, advisories, and warnings?" (if so, they should be specified) and "Have any safety procedures and environmental guidelines been included in operational plans?" (if so, descriptions of them should be furnished).[135]

Over half of the state weather modification control acts require cloud seeders to file reports about their activities. Although some of the laws demand little, others do require a significant amount of environmental information,[136] as in Colorado.[137]

Reporting requirements do not ensure compliance, but potential use of the information by the government does. If a report to NOAA indicates that a project significantly departs from the practices generally used in similar snowpack operations to avoid dangers such as avalanches, flooding, and other environmentally harmful consequences, NOAA will notify both the project operator and the responsible state officials. The agency includes recommendations where appropriate.[138] Such regulation by adverse publicity can be most effective.[139]

3. Banning snowpack augmentation

Many target areas in the Colorado River Basin for snowpack augmentation are federal lands that have been or may be designated as wilderness areas. The Wilderness Act of 1964 and subsequent laws creating specific wilderness areas form a system of lands legislatively preserved and protected in their natural condition. These areas are to "be administered for the use and enjoyment of the American people in such manner as will leave them unimpaired for future use and enjoyment as wilderness." The basic law states:

> A wilderness, in contrast with those areas where man and his own works dominate the landscape, is . . . an area where the earth and its community of life are untrammeled by man . . . retaining its primeval character and influence, without permanent improvements or human habitation, which is protected and managed so as to preserve its natural condition, and which . . . generally appears to have been affected primarily by forces of nature, with the imprint of man's work substantially unnoticeable, . . . of sufficient size to make practicable its preservation and use in an unimpaired condition.[140]

The question arises whether seeding from the ground in wilderness areas or releasing materials over them or upwind from them results in unnatural conditions incompatible with the intent of Congress. In at least one lawsuit, involving efforts to increase snowpack above the reservoir created by Hungry Horse Dam in Montana, such arguments were advanced.[141] Because the government sponsors of the project decided they had all the snow they wanted for the season and determined not to seed, the case was mooted and did not come to trial.[142]

The Bureau of Reclamation's policy is that effects of cloud seeding "are not manifested as an observable artificiality in wilderness character and that the Wilderness Act was not intended to, and does not, prohibit weather modification."[143] Some units of the National Park Service and of the U.S. Forest Service interpret the law differently[144] and assert that artificial snowpack efforts result in unnatural conditions incompatible with congressional intent. These agencies restrict installation and use of hydrometeorological data collection equipment in wilderness areas.[145] As long as proposed legislation resolving these differences between the agencies and their local representatives is not enacted[146] or litigation does not result, the legal uncertainty respecting this aspect of snowpack enhancement will continue.

Tropical Storm Modification

Scenario

The average annual damage in the United States over a 10-year period from hurricanes exceeds $450 million. Most of the losses result from the force of the wind and the storm surge. Past experiments of Project Stormfury, operated mainly by the Department of Defense and NOAA, suggest that maximum wind speeds can be decreased and that this in turn will reduce the storm surge. The current planning document for Stormfury-Americas, 1976-1978, notes that "if we can learn to reduce the maximum surface wind speeds by even only 10 to 15 percent, the hurricane damage in the United States could be reduced by $50 to $100 million per year, and lives might be saved in difficult-to-evacuate areas."[147]

The areas in which hurricane seeding experiments would be performed are the western North Atlantic, the Gulf of Mexico, and the Caribbean Sea. Experiments in these areas are conducted over the high seas, but a tropical storm is eligible for seeding only if there is a 10% probability or less of the hurricane center coming within 50 nautical miles (93 km) of a populated land area within 24 hours after seeding.[148] This precaution is motivated by at least two technical reasons: (a) a storm treated well out at sea will revert to its natural state before reaching land, and (b) changes occur in a hurricane's circulation pattern when portions of it pass over land, and these land-induced modifications might complicate evaluation of results produced by seeding.[149]

There are, of course, other reasons for caution in selecting sites for seeding. Because Project Stormfury is an experimental program, there is much to be learned about seeding cyclonic storms. Should modification efforts reduce both wind and rain associated with tropical storms, harm as well as benefits might result. For example, in 1966 Hurricane Inez threatened the Gulf Coast of the United States but produced enough precipitation to fill the reservoirs of the Mexican altiplano. A substantial portion of the runoff from Mexico in the Lower Rio Grande River comes from tropical storms.[150] Another reason for seeding storms only far at sea is the possibility that whatever harm might result from a seeded hurricane would be blamed upon the United States. Enough of the world's troubles are laid at our doorstep without courting further difficulties. Since nations other than the United States might be influenced by Project Stormfury, questions of transnational law, rather than domestic law only, must be considered.[151] At least four major issues emerge.

1. Notification of hurricane modification projects

Must the United States give notice to potentially affected countries? Assuming that some notice must be afforded, what should be its timing and substance?

(a) *Need to give notice.* Nations are obligated by customary international law to warn other nations of dangerous conditions of which they are aware. The International Court of Justice set forth this principle in the *Corfu Channel* case. The Albanian government, aware of a minefield in the Corfu Channel off its coast, failed to warn a British squadron of warships which it knew was entering the mined area. Exploding mines damaged two ships, and three weeks later the British undertook to clear the channel of mines, including ones in Albanian territorial waters. The British, concerned about damage to their vessels, and the Albanians, angered about the intrusion into their sovereignty, took their dispute to the Security Council of the United Nations, which referred it to the International Court of Justice. The court noted:

> The obligations incumbent upon the Albanian authorities consisted in notifying, for the benefit of shipping in general, the existence of a minefield in Albanian territorial waters and in warning the approaching British warships of the imminent danger .... Such obligations are based ... on certain general and well-recognized principles, namely: elementary considerations of humanity, even more exacting in peace than in war; the principle of freedom of maritime communication; and every State's obligation not to allow knowingly its territory to be used for acts contrary to the rights of other States. ... In fact nothing was attempted by the Albanian authorities to prevent the disaster. These grave omissions involve the international responsibility of Albania.[152]

Corfu Channel requires a nation that knows of a dangerous condition to notify others that might be harmed by it.

The United States and Great Britain complied with this obligation prior to conducting tests of nuclear devices in the Pacific[153] before the nuclear test moratorium[154] and later when testing such devices in the atmosphere.[155] They in effect staked out a portion of the high seas for a limited time and told persons entering it that they did so at their own risk. Japanese protests[156] were answered with notes indicating that the tests would proceed[157] and they did.

One of the provisions of the Antarctic Treaty—to which the United States is a party—requires advance notice of all expeditions to and within Antarctica.[158] This is in accord with the time-honored practice of giving notice of naval maneuvers and the like.

When initiating Project Stormfury, the United States realized that it might affect other nations. From 1967 through 1971, the Department of State sent telegrams to the Caribbean countries and other nations notifying them of our intent to conduct seeding operations.[159] Also, from 1968 through 1972, representatives of the United States met with those of foreign countries in hurricane-prone areas to discuss plans for Project Stormfury and related subjects.[160]

At an informal meeting in November 1975, scientists and lawyers representing various member nations of the United Nations Environment Programme (UNEP) and the World Meteorological Organization (WMO) agreed on the following principle on notification:

> States shall in good faith give adequate and timely notification of

prospective major weather modification activities, within their jurisdiction or control, to WMO which should transmit such notification to all interested States.[161]

This view, which is of course not binding in itself,[162] would seem to satisfy our obligations with respect to Project Stormfury activities.

(b) *Timing and substance of notice.* What is timely? What is adequate notice? An informal executive agreement between the United States and Canada provides for transmission of information related to weather modification activities "as soon as practicable" and, if possible, "prior to the commencement of such activities." In any event, pertinent information should be sent within five working days.[163]

Both nations have reported to each other under this agreement,[164] but none of the activities is as significant as Project Stormfury, and the substance of information is relatively light. More extensive information is provided by the WMO register of national weather modification activities,[165] to which the United States plans to send reports on 73 projects conducted in 1975.[166] This, of course, is after-the-fact reporting rather than advance notification as suggested by the UNEP-WMO informal meeting. Sending something more than a Department of State telegram and news releases is desirable, and advance availability of planning documents for Project Stormfury would be in order.

2. Consultation with potentially affected nations

Once a nation is notified of planned hurricane seeding, is the United States obligated to consult with that country about it? Our notices to date have not elicited much interest by other nations that might be affected by the hurricane seeding,[167] perhaps because they will gain, not lose, if the modification effort is successful.[168]

But, as noted above, the Japanese did object to nuclear testing,[169] and both the British and the Americans consulted with them and replied to their objections.[170] The Space Treaties contain mandatory consultation clauses.[171] These actions promote international understanding and cooperation and help avoid tensions and disputes. Environmental protection is clearly a matter on which consultation is required.

Legal uncertainty again is an issue. Participants at the UNEP-WMO informal meeting did not draft a principle that would compel states to consult but did recommend the following:

It is desirable that a State, in whose territory major weather modification activities are to be undertaken, should engage in meaningful and timely consultation with interested states at their request, with a view to working out mutually acceptable arrangements regarding the conduct of those activities.[172]

The one case most frequently cited as applying to this matter is *Lake*

Lanoux. In that arbitration between Spain and France over a French plan to divert waters from a stream shared by two nations and then to replace it with water of the same quality and quantity, the tribunal noted:

> As a matter of form, the upstream State has procedurally a right of initiative; it is not obliged to associate the downstream State in the elaboration of its projects. If, in the course of discussions, the downstream State submits projects to it, the upstream State must examine them, but it has the right to give preference to the solution contained in its own project, provided it takes into consideration in a reasonable manner the interests of the downstream State.[173]

One reasonable conclusion applicable to Project Stormfury is that the United States, after giving notice to other nations, should assess any information, suggestions, and objections those nations might offer, but not necessarily to the extent of undermining the project.

3. Finality of decision to attempt modification

The *Lake Lanoux* case stands for the proposition that no potentially affected nation has veto power over any seeding during Project Stormfury. One author explained:

> Forcible measures can exceptionally be taken on foreign territory when forces of nature in that territory cause a natural catastrophe in the neighboring state, for example, floods or fire across the border. Here the affected state can use force only when action of the local authorities proves inadequate and does not succeed in containing the dangers.[174]

So, unless the nations that object to our efforts to tame the destructive power of hurricanes wish to act themselves to protect us, we are entitled to seed those cyclonic storms in the Project Stormfury area that would endanger the United States.[175] The power to decide whether to seed rests with the United States. Of course, many of the storms in the Project Stormfury area will not threaten American interests.

The following proposition, submitted to the General Assembly Committee on Peaceful Uses of the Sea-Bed and the Ocean Floor Beyond the Limits of National Jurisdiction, would encourage experimental seeding:

> Scientific research in the international seas shall be open to any State and shall be promoted and facilitated under forms of co-operation and assistance which permit the participation of all States, irrespective of their level of development or of whether they are coastal or landlocked.[176]

After the Test Ban Treaty, which France did not sign but which so many nations did that it might be regarded as binding under customary international law, France announced plans to test atomic devices in the South Pacific. The tests proceeded even though New Zealand and Australia argued forcefully before the International Court of Justice.[177] A continuing uncertainty is whether a nation might regard an effort to prevent it from undertaking a planned activity to be an infringement of its sovereignty.

4. Liability for damages in other nations

Several international law cases indicate that states causing harm to other nations have an obligation to compensate them for the losses. An example is the *Trail Smelter* arbitration between the United States and Canada, which arose when fumes from smelters in British Columbia damaged farms in Washington.[178] The persons who suffered the harm could prove causation. As indicated in the discussion about liability of cloud seeders,[179] the party seeking relief must prove the causal linkage between the activity of the defendant and the loss of the plaintiff. This will be an almost insuperable problem for nations asserting that hurricane seeding caused damage. Perhaps this will be easier once the scientific uncertainties are resolved.

Assuming that a nation could establish causation, it then could act to strengthen its case. A reservation of rights, such as the Japanese expressed during the nuclear testing,[180] would support a later argument that they had not estopped[181] themselves from obtaining compensation by acquiescing in the conduct of the modifiers.[182]

It might be argued that the Canadian-American agreement for reporting modification activities manifests our recognition of the potentially harmful consequences of weather modification and that we should be estopped to deny that recognition. A proper reply would be that the agreement by its own terms is not to be extended to cover other situations.[183]

The so-called Helsinki Rules, adopted by the nongovernmental International Law Association in 1966, sets forth a "reasonable man" test for apportioning water in international drainage basins. Some nations apparently have merged these rules with their normal practices.[184] Under this test, a nation that deprived another nation of precipitation by seeding a hurricane would be liable only if the modification effort could be regarded as inconsistent with the legal notion of reasonableness, assuming, of course, that the action did reduce precipitation.

Finally, Principles 9, 21, and 22 of the Stockholm Declaration on the Human Environment deserve mention, as noted by one recent discussion of them:

> *Principle 9.* Environmental deficiencies generated by the conditions
> of underdevelopment and natural disasters posed grave problems

and could best be remedied by accelerated development through the transfer of . . . technological assistance . . . .

Principle 21.. States had . . . the sovereign right to exploit their own resources pursuant to their own environmental policies, and the responsibility to ensure that activities within their jurisdiction or control did not cause damage to the environment of other States or of areas beyond the limits of national jurisdiction.

Principle 22. States were to co-operate to develop further the international law regarding liability and compensation for the victims of pollution and other environmental damage caused by activities within the jurisdiction or control of such States to areas beyond their jurisdiction.[185]

Nations with the ability to seed hurricanes may have difficulty in adhering to all of these principles. They are expected to solve problems of natural disasters and to cooperate with other countries, but to avoid causing harm while doing so. They may not be able to act without causing damage.

Conclusion

The preceding scenarios should demonstrate that we still encounter serious uncertainties about the law of weather modification. Perhaps lawyers should raise their cups in a toast to the "glorious uncertainty of the law," for it is from our participation in the decision-making process that we make our living and promote sounder rules for technological activities.

Acknowledgments

Research assistants who participated in studies upon which much of this paper is based are Steven Cox, Guy Fletcher, Steven Hernandez, Patricia Sterns, and Tanis Toll. Portions of those studies were funded by grants from or contracts with the Office of Water Research and Technology, National Science Foundation, and the National Oceanic and Atmospheric Administration.

Notes

1. Charles Macklin, Love a la Mode, act ii, scene 1, *in* J. O. Bartley, ed., Four Comedies by Charles Macklin (1968).
2. Panama Ref. Co. v. Ryan, 293 U.S. 388 (1935).

3. For discussion of the argument in the case, see Jaffe & Nathanson, Administrative Law: Cases and Materials 61 (2d ed. 1961).

4. Epictetus, Discourses, bk. I, ch. 27.

5. Tucson Daily Citizen, May 28, 1976, at 7, col. 7.

6. 1969 Minn. Laws ch. 771.

7. Minn. Stat. Ann. sec. 645.021 (West Supp. 1975).

8. The counties which did file were: Big Stone County, 1971 Minn. Laws sec. 2404; Chippewa County, 1971 Minn. Laws sec. 2405; Grant County, 1971 Minn. Laws sec. 2407; Lac Qui Parle County, 1969 Minn. Laws sec. 2791; and Yellow Medicine County, 1969 Minn. Laws sec. 2799. For discussion of the Minnesota experience, see Davis, Weather Modification Law Developments, 27 Okla. L. Rev. 409, 424 (1974).

9. Letter from Thomas Kalitowski, Assistant Commissioner of Agriculture, State of Minnesota, to Ray Jay Davis, Dec. 19, 1975; and interview with Randy Young, Minnesota Department of Agriculture at Kansas City, Mo., Jan. 15, 1976.

10. See text at notes 125-46 infra.

11. See text at notes 57-60 infra.

12. The material presented here is from a study prepared for the Technology Assessment of Suppression of Hail (TASH) program funded by the RANN Program of the National Science Foundation.

13. See, e.g., Ariz. Rev. Stat. Ann. sec. 45-2401 (Supp. 1975-76).

14. See, e.g., S.D. Comp. Laws sec. 38-9-4-1 (Supp. 1975).

15. Wash. Rev. Code Ann. sec. 43.37.010 (1970), as amended (Supp. 1975).

16. Mass. Ann. Laws ch. 6, 17 (Supp. 1974).

17. See, e.g., Neb. Rev. Stat. sec. 2-2403 (1970).

18. As illustrations, the Aeronautics Commission in North Dakota is the umbrella agency within which weather control officials are housed. N.D. Cent. Code sec. 2-07-02.3 (1975). The Wisconsin Public Utilities Commission possesses what little regulatory authority there is in that state. Wis. Stat. Ann. sec. 195.40 (1957), as amended (Supp. 1975-76). And in Illinois the State Weather Modification Board is within the Department of Registration and Education. Ill. Stat. Ann. ch. 146¾, sec. 3.01 (Smith-Hurd Supp. 1976-77).

19. E.g., Connecticut. Conn. Gen. Stat. Ann. sec. 24-5 (1975).

20. E.g., Colorado. Colo. Rev. Stat. Ann. sec. 36-20-107 (1973).

21. E.g., Illinois. Ill. Stat. Ann. ch. 146¾, sec. 4(b) (Smith-Hurd Supp. 1976-77).

22. Colorado has a fairly specific listing of criteria in its weather control law. Colo. Rev. Stat. Ann. sec. 36-20-107 (1973). Wyoming at one time permitted licensing as weather modifiers only persons who held valid licenses as registered professional engineers. Wyo. Stat. Ann sec. 9-271 (Supp. 1971).

This provision has now been deleted by amendment. Wyo. Stat. Ann sec. 9-271 (Supp. 1975).

23. *See* W. Gellhorn, Individual Freedom and Governmental Restraint 109 (1956).

24. The Idaho law merely requires registration of a project, Idaho Code sec. 22-3201 (1968), contrasted with the Texas permit provision, Texas Water Code sec. 14.061 (1972), *as amended* (Supp. 1975).

25. For an illustration of the type of information a modern regulatory agency demands, see Appendix III to the Texas Water Development Board, Rules, Regulations and Modes of Procedure Relating to the Texas Weather Modification Act (1976).

26. *See, e.g.,* Utah Rules, Regulations and Procedures ch. 5 (1973).

27. For illustrative provisions, see the Illinois law, Ill. Stat. Ann. ch. 146¾, sec. 21 (modification), sec. 23 (suspension and revocation) (Smith-Hurd Supp. 1976-77).

28. *See, e.g.,* Cal. Water Code sec. 409 (West 1971).

29. Ill. Stat. Ann. ch. 146¾, sec. 17(c) (Smith-Hurd Supp. 1976-77).

30. *See, e.g.,* Mass. Ann. Laws ch. 6, sec. 72 (1962).

31. *See, e.g.,* Mont. Rev. Codes Ann. sec. 89-318 (Supp. 1975).

32. 15 U.S.C. sec. 330 (Supp. VI 1976).

33. 15 C.F.R. sec. 908 (1976).

34. *See, e.g.,* Texas Water Development Board, *supra* note 25, ch. 6, rule 600.2(b); Utah Rules, *supra* note 26, ch. 9.

35. Federal expenditures have been detailed in reports of the International Committee on Atmospheric Sciences that were issued at least annually during the 1960s and until NOAA assumed the federal role of receiving weather modification reports in the 1970s. For one of the most detailed accounts of actual and recommended federal spending, see Interdepartmental Commission for Atmospheric Sciences, Program: Fiscal Year 1969, Rep. No. 12 (1968). For a discussion of federal weather modification projects and their legal setting, see I. Gutmanis & R. Gillis, Weather Modification: Programs and Prospects, 2 Envt'l Rep. Monograph 8 (1971).

36. N.D. Cent. Code sec. 2-07-02.5(4) (1975).

37. S.D. Comp. Laws sec. 38-9-9 (Supp. 1975). An appropriations bill, S.B. 208, which would have provided funding for the South Dakota program, failed in 1976. See Donnan, Pellett, Leblang, & Ritter, The Rise and Fall of the South Dakota Weather Modification Program, 8 J. Weather Modification 1 (1976).

38. Utah Code Ann. sec. 73-15-3 (Supp. 1975); 1975 Utah Laws, ch. 128, sec. 1; 1976 Utah Laws ch. 37, item 144.

39. The Supreme Court noted that people who do business with the

government will usually "turn square corners." Rock Island, A. & L.R.R. v. United States, 254 U.S. 141, 143 (1920).

It is worth noting that during the period after Congress terminated the power of the National Science Foundation to require weather modification activity reporting (Pub. L. No. 90-407, sec. 11, 82 Stat. 360 (1968)), most of the modifiers who voluntarily continued reporting to the foundation were persons who held federal weather modification contracts. For discussion of contractual control over cloud seeding, see Davis, State Regulation of Weather Modification, 12 Ariz. L. Rev. 35, 60-62 (1970).

40. *See, e.g.,* N.D. Cent. Code sec. 2-07-11 (1975).

41. *See, e.g.,* Cal. Gov't Code sec. 53063 (West 1966).

42. *See, e.g.,* Iowa Code Ann. sec. 361.1 (Supp. 1976).

43. N.D. Cent. Code sec. 2-07-11 (state), sec. 2-07-11.1 (county) (1975); S.D. Comp. Laws sec. 38-9-11 (state) (1967), sec. 38-9-11.1 (county) (Supp. (1975).

44. *See, e.g.,* Okla. Stat. Ann. tit. 2, sec. 1423 (1973).

45. *See* discussion in text at notes 6-9 *supra.*

46. *See, e.g.,* 1975 Laws of South Dakota, ch. 354.

47. K. Browning & B. Foote, Airflow and Hail Growth in Supercell Storms and Some Implications for Hail Suppression, NHRE Tech. Rep. 51/1 (May 1975).

48. Negative comments about alleged weather modification activities were reported to Congress in Hearings on Progress in Weather Modification Before Subcomm. on Water and Power Resources of the Senate Comm. on Interior and Insular Affairs, 90th Cong., 1st Sess. (1967).

49. Pennsylvania *ex rel.* Township of Ayr v. Fulk, No. 53 (C.P. Fulton County, Pa., Feb. 28, 1968).

50. *See* Md. Ann Code art. 66C, sec. 110A (1970). This provision was enacted by 1965 Md. Laws, ch. 192, which banned weather modification from March 30, 1965, until Sept. 1, 1967, and 1968 Md. laws ch. 587 extended the ban from July 1, 1968, until Sept. 1, 1969; and 1969 Md. Laws ch. 22 further extended it from July 1, 1969, to Sept. 1, 1971. No further ban has been passed. 1973 Md. Laws 1st Sp. Sess., ch. 6, sec. 2 repealed the law.

51. Pa. Stat. Ann. tit. 3, sec. 1114 (Supp. 1975-76).

52. W. Va. Code Ann. sec. 29-2B-13 (1971).

53. For discussion of weather modification litigation through early 1974, see Davis, *supra* note 8, at 412-15.

54. Southwest Weather Research, Inc. v. Rounsaville, 320 S.W.2d 211, and Southwest Weather Research, Inc. v. Duncan, 319 S.W.2d 940 (Tex. Civ. App. 1958), *both aff'd sub. nom.* Southwest Weather Research, Inc. v. Jones, 160 Tex. 104, 327 S.W.2d 417 (1959); Farmers and Ranchers for Natural Weather

v. Atmospherics, Inc., Civ. No. 7594 (D. Ct. Lamb County, Tex., May 3, 1974); Atmospherics, Inc. v. Ten Eyck, Civ. A. (D. Ct. Alamosa County, Colo., April 4, 1973); Shawcroft v. Department of Natural Resources, Civ. A. (D. Ct. Alamosa County, Colo., Sept. 20, 1972); Pennsylvania *ex rel.* Township of Ayr v. Fulk, No. 53 (C.P. Fulton County, Pa., Feb. 28, 1968); Pennsylvania Natural Weather Ass'n v. Blue Ridge Weather Modification Ass'n, 44 Pa. D. & C. 2d 749. (C.P. Fulton County, Pa., Feb. 28, 1968); Auvil Orchard Co. v. Weather Modification, Inc., No. 19268 (Super. Ct. Chelan County, Wash., 1956).

55. Pennsylvania *ex rel.* Township of Ayr v. Fulk, No. 53, (C.P. Fulton County, Pa., Feb. 28, 1968); Pennsylvania Natural Weather Modification Ass'n v. Blue Ridge Weather Modification Ass'n, 44 Pa. D. & C. 2d 749 (C.P. Fulton County, Pa., Feb. 28, 1968).

56. Pennsylvania *ex rel.* Township of Ayr v. Fulk, No. 53 (C.P. Fulton County, Pa., Feb. 28, 1968).

57. Southwest Weather Research, Inc. v. Rounsaville, 320 S.W.2d 211, and Southwest Weather Research, Inc. v. Duncan, 319 S.W.2d 940 (Tex. Civ. App. 1958), *both aff'd sub nom.* Southwest Weather Research, Inc. v. Jones, 160 Tex. 104, 327 S.W.2d 417 (1959).

58. Davis, Book Review, 48 S. Cal. L. Rev. 580 (1974).

59. *See* Ver Hagin v. Gibbons, 47 Wis. 2d 220, 177 N.W.2d 83 (1970); *but see* Barci v. Intalco Aluminum Corp., 11 Wash. App. 342, 522 P.2d 1159 (1974).

One of the problems in entering judgments for intangible injuries is that they are so differently perceived by each person. This calls to mind the limerick reported in L. Untermeyer, Lots of Limericks: Light, Lusty and Lasting 53 (1961):

> There was a faith-healer of Deal
> Who said, "Although pain isn't real,
> If I sit on a pin,
> And it punctures my skin,
> I dislike what I fancy I feel."

60. W. Prosser, Law of Torts sec. 54 (4th ed. 1971).

61. Southwest Weather Research, Inc. v. Rounsaville, 320 S.W.2d 211, and Southwest Weather Research, Inc. v. Duncan 319 S.W.2d 940 (Tex. Civ. App. 1958), *both aff'd sub. nom.* Southwest Weather Research, Inc. v. Jones, 160 Tex. 104, 327 S.W.2d 417 (1959).

62. Slutsky v. City of New York, 197 Misc. 730, 97 N.Y.S.2d 238 (Sup. Ct. 1950).

63. Pennsylvania Natural Weather Ass'n v. Blue Ridge Weather Modification Ass'n, 44 Pa. D. & C. 2d 749 (C.P. Fulton County, Pa., Feb. 2, 1968).

64. Colo. Rev. Stat. Ann. sec. 36-20-103 (1973).

65. Utah Code Ann. sec. 73-15-4 (Supp. 1975).

66. N.D. Cent. Code sec. 2-07-01 (1975). State statutes concerning atmospheric water rights are analyzed in Fischer, Weather Modification and the Right of Capture, 8 Nat. Resources Law. 639 (1976).

67. City of Eveleth v. Ruble, 225 N.W.2d 521 (1974); Allen, Liability of Architects and Engineers to Third Parties, 22 Ark. L. Rev. 454 (1968).

68. Covil v. Robert & Co., Assoc. 112 Ga. App. 163, 144 S.E.2d 450 (1965); Miller v. Raaen, 273 Minn. 109, 139 N.W.2d 877 (1966); City of Eveleth v. Ruble, 225 N.W.2d 521 (1974).

69. Mann, The Yuba City Flood: A Case Study of Weather Modification Litigation, 46 Bull. Am. Meteor. Soc'y 690 (1968).

70. W. Va. Code Ann. sec. 29-2B-13 (1971).

71. Pa. Stat. Ann. tit. 3, sec. 1114 (Supp. 1975-76).

72. Restatement (Second) of Torts sec. 520(b), (c) (Tent. Draft No. 10, 1964).

73. See, e.g., Ill. Ann. Stat. ch. 146¾, sec. 28(a) (Smith-Hurd Supp. 1976-77); Tex. Water Code sec. 14.102(a) (1972); N.D. Cent. Code sec. 2-07-10.1 (1975).

74. Prosser, Law of Torts sec. 87 (4th ed. 1971).

75. Slutsky v. City of New York, 197 Misc. 730, 97 N.Y.S.2d 238 (Sup. Ct. 1950).

76. Pennsylvania Natural Weather Ass'n v. Blue Ridge Weather Modification Ass'n, 44 Pa. D. & C. 2d 749 (1968).

77. Southwest Weather Research, Inc. v. Rounsaville, 320 S.W.2d 211, and Southwest Weather Research, Inc. v. Duncan, 319 S.W.2d 940 (Tex. Civ. App. 1958), both aff'd sub. nom. Southwest Weather Research, Inc. v. Jones, 160 Tex. 104, 327 S.W.2d 417 (1959).

78. Farmers and Ranchers for Natural Weather v. Atmospherics, Inc., Civ. No. 7594 (D. Ct. Lamb County, Tex., May 3, 1974).

79. Contributory negligence consists of conduct on the part of the plaintiff that falls below that which a reasonable person would exercise in care of his or her own welfare. Restatement (Second) of Torts sec. 463 (1965).

80. "Consent" denotes willingness that an invasion of an interest take place. Restatement (Second) of Torts sec. 10A (1965).

81. Haterkamp v. City of Rock Hill, 316 S.W.2d 620 (1958). See also 93 C.J.S. Water sec. 114 (1956).

82. Prosser, Law of Torts sec. 131 (4th ed. 1971).

83. 28 U.S.C. sec. 2680(a) (1975); Reynolds, The Discretionary Function Exception of the Federal Tort Claims Act, 57 Geo. L.J. 81 (1968); Reynolds, Strict Liability Under the Federal Tort Claims Act: Does Wrongful Cover a Few Sins, No Sins, or Non-Sins? 23 Am. U.L. Rev. 813 (1974).

84. Dalehite v. United States, 346 U.S. 15 (1953); Bartie v. United States,

316 F.2d 754 (5th Cir. 1964); National Mfg. Co. v. United States, 210 F.2d 263 (8th Cir. 1954).

85. Swanson v. United States, 229 F. Supp. 217 (N.D. Cal. 1964).

86. Fed. Rules Civ. Proc., Rule 23; 28 U.S.C. 289 (1972).

87. Weeks v. Bareco Oil Co., 125 F.2d 84 (7th Cir. 1941); Daer v. Yellow Cab Co., 67 Cal. 2d 695, 433 P.2d 732 (1967).

88. Eisen v. Carlisle & Jacquelin, 417 U.S. 156, (1974); Zahn v. International Paper Co., 414 U.S. 291 (1973); *cf.* Alyseka Pipeline Service Co. v. Wilderness Soc'y, 421 U.S. 240 (1975).

89. *See* note 12 *supra.*

90. For discussion of the background of the snowpack augmentation possibilities and some of their legal implications, see R. Davis, United States and Mexico: Weather Technology, Water Resources and International Law, 12 Nat. Resources J. 530, 532-36 (1972).

91. This organization of state governments annually publishes a volume containing the texts of various proposed statutes that might be enacted by the various states. See, *e.g.,* 35 Council of State Governments, Suggested State Legislation (1976).

92. For discussion of the possibility of uniformity through this route, see R. Davis, Uniformity Among Weather Modification Laws 11-13, ASCE National Convention Preprint 2548 (Nov. 1975).

93. Utah Code Ann. sec. 73-15-8 (Supp. 1975).

94. Colo. Rev. Stat. Ann. sec. 36-20-118 (1973).

95. N.M. Stat. Ann. sec. 75-37-12 (1968).

96. Vineyard Co. v. Twin Falls Co., 245 F. 9 (9th Cir. 1917).

97. Wyoming v. Colorado, 298 U.S. 573 (1936).

98. Kansas v. Colorado, 206 U.S. 46 (1907); see also Brooks v. United States, 119 F.2d 636 (9th Cir. 1941).

99. Nebraska v. Wyoming, 325 U.S. 589 (1945).

100. 373 U.S. 546 (1963).

101. Haber, Arizona v. California—a Brief Review, 4 Nat. Resources J. 17 (1964).

102. Clyde, The Colorado River Decision—1963, 8 Utah L. Rev. 299, 312 (1964).

103. *E.g.,* Tennessee Valley Authority, 16 U.S.C. secs. 831 *et seq.* (1971).

104. Domestic Council, The Federal Role in Weather Modification (1975).

105. U.S. Const. art. I, sec. 8, cl. 3.

106. Gibbons v. Ogden, 22 U.S. (9 Wheat.) 1 (1824).

107. 40 C.F.R. 125 (1975) (Environmental Protection Agency Regulations).

108. R. Davis, The Legal Implications of Atmospheric Water Resources Development and Management 100-101 (1968).

109. Tobin, The Interstate District and Cooperative Federalism, 36 Tulane L. Rev. 67 (1961).

110. U.S. Const. art. 1, sec. 10.

111. Landes v. Landes, 153 N.Y.S.2d 14 (1956), *appeal dismissed,* 352 U.S. 948 (1956).

112. *See* By-Laws of the North American Interstate Weather Modification Council, art. III, *as amended* (1976).

113. Memorandum of Understanding Between the States of Kansas, Nebraska, and Colorado and the U.S. Department of the Interior, Bureau of Reclamation, for Cooperative Weather Research in the High Plains (1974).

114. Mewes, Hail Suppression in Southern Idaho, HERS TASH Working Paper No. 20 (1976); W. Jenkins, Utah-Idaho Weather Modification Corporation 66, 73, Cloud Seeding Seminar, Utah Department of Natural Resources, Division of Water Resources, Salt Lake City, Utah, 1975; Utah Division of Water Resources, 2 Newsletter Weather Modification in Utah (1976, No. 1). Utah Division of Water Resources, Newsletter, Weather Modification in Utah (May 19, 1976).

115. *Cf.* Davis, *supra* note 39, at 67-69.

For cases relating to interstate natural resources compacts, see League to Save Lake Tahoe v. Tahoe Regional Planning Agency, 507 F.2d 517 (9th Cir. 1974), *cert. denied,* 95 S. Ct. 1398 (1975); Borough of Morrisville v. Delaware River Basin Comm'n, 399 F. Supp. 469 (E.D. Pa. 1975).

116. U.S. Const. art. I, sec. 10, cl. 3.

117. 341 U.S. 22 (1951).

118. Anthony v. Veatch, 220 P.2d 493 (Ore. 1950).

119. State v. Knapp. 167 Kan. 546, 207 P.2d 440 (1949).

120. 39 N.J. Super. 33, 120 A.2d 504 (1956).

121. Weakley, Interstate Compacts, 3 Nat. Resources L. 81, 87 (1970).

122. National Water Commission, Water Policies for the Future 418-24 (1973).

123. 75 Stat. 688 (1961).

124. The DRBC has, in spite of its powers, proved disappointing to some informed commentators. Ackerman, Ackerman, Sawyer & Henderson, The Uncertain Search for Environmental Quality (1974).

125. *See, e.g.,* Wilkins, No One Owns the Rain, Field & Stream, Mar. 1976, at 30-31, 122-25.

126. For a full-fledged technology assessment of augmentation of snow-pack in the Colorado River Basin, see Weisbecker, The Impacts of Snow Enhancement: Technology Assessment of Winter Orographic Snowpact Augmentation in the Upper Colorado River Basin (1974).

127. 42 U.S.C. sec. 4332 (C) (1970).

128. The cases decided by 1973 are discussed in F. Anderson, NEPA in

the Courts: A Legal Analysis of the National Environmental Policy Act (1973).

129. *See* 40 C.F.R. sec. 6(gg) (1975), 40 Fed. Reg. 16814 (April 14, 1975).

130. *See* Davis, Legal Response to Environmental Concerns about Weather Modification, 14 J. Appl. Meteor. 681 (1975); Hagman, NEPA's Progeny Inhabit the State—Were the Genes Defective? 7 Urban L.J. 3 (1974); Yost, NEPA's Progeny: State Environmental Policy Acts, 3 Envt'l L. Rep. 50090 (1973).

131. 1973 N.M. Laws ch. 310.

132. 1974 N.M. Laws ch. 46. Comment, The Rise and Demise of the New Mexico Environmental Quality Act, "Little NEPA," 14 Nat. Resources J. 401 (1974).

133. 15 U.S.C. secs. 330-330e (Supp. VI 1976).

134. 37 Fed. Reg. 22974 (1972).

135. 15 C.F.R. sec. 908.4(a) (1976).

136. *See* discussion in Davis, *supra* note 8, at 436-37.

137. Colo. Dep't Nat. Res., Weather Mod. Rules & Regs., Rule II (G), (H) (1973).

138. 15 C.F.R. sec. 908.12(d) (1976). M. Charak & M. Digiulian, Weather Modification Activity Reports: Nov. 1, 1972-Dec. 31, 1973, NOAA, at D-1, D-2 (1974).

139. Gellhorn, Adverse Publicity by Administrative Agencies, 86 Harv. L. Rev. 1380 (1973).

140. 16 U.S.C. sec. 1131(c) (Supp. IV 1974).

141. Montana Wilderness Ass'n v. Hodel, Civ. No. 74-5-GF (D. Mont., Jan. 30, 1974).

142. There has been no snowpack augmentation area in the drainage basin behind the dam since the litigation was filed. It would appear that the plaintiffs won their point without the need to go to trial.

143. Division of Atmospheric Water Resources Management, Position Paper on Weather Modification Over Wilderness Areas and Other Conservation Areas (Introductory abstract, Mar. 1974).

144. *See* Foote, Wilderness—a Question of Purity, 3 Envt'l L. 225 (1973).

145. *E.g.,* Letter from Thomas Nelson, Deputy Chief, National Forest Service, to Warren Johm, Aug. 14, 1973.

146. H.R. 12316, 93d Cong., 2d Sess (1974) (Sisk bill), would allow cloud seeding and use of data collection instrumentation in wilderness areas, if the departments administering them determined there would be no adverse impact.

147. Environmental Research Laboratories, Program Development Plan Stormfury-Americas 1976-78 (Oct. 10, 1975).

148. Interview with Merlin Williams, Program Director for Weather Modification, NOAA (Feb. 27, 1976).

149. Environmental Research Laboratories, Project Stormfury Annual Report (1972).

150. Roberts, We're Doing Something About the Weather, 141 Nat'l Geographic 518, 545 (Apr. 1972).

151. International legal implications of weather modification are reported in Hassett, Weather Modification and Control: International Organizational Prospects, 7 Tex. Int. L.J. 89 (1971), and in Samuels, International Control of Weather Modification Activities: Peril or Policy? 13 Nat. Resources J. 327 (1973).

152. Corfu Channel Case, 43 A.J.I.L. 558, 570-71 (1949).

153. 4 Whiteman, Digest of International Law 583, 597 (1965).

154. Id. at 601.

155. Nuclear Test Ban Treaty, TIAS no. 5344; 480 U.N.T.S. 43, 57 AJIL 1026 (1963); State Department Bull. No. 1259 at 239 (1963).

156. 4 Whiteman, Digest of International Law 585, 596 (1965).

157. Id. at 586, 596.

158. The Antarctic Treaty, Dec. 1, 1959 [1961], 12 U.S.T. 794, 402 U.N.T.S. 71 (effective June 23, 1961).

159. Memorandum to Merlin Williams, Program Director for Weather Modification, NOAA, from William Mallinger, Assistant Manager for Field Research Operations, NOAA (Feb. 13, 1976).

160. Id.

161. WMO and UNEP, Report of the WMO/UNEP Informal Meeting on Legal Aspects of Weather Modification, Geneva, Nov. 17-21, 1975, at 6.

162. Before this would become binding, the WMO Executive Committee would first refer it to the WMO Congress, which then could approve it. The final resolution would be binding on all member states of the WMO. For all U.N. member states to become bound, the U.N. General Assembly would have to approve it.

163. Agreement Between the USA and Canada Relating to the Exchange of Information on Weather Modification Activities (26 Mar. 1975). Reprinted at 1 Envt'l L. & Pol'y 109 (1975).

164. E.g., Atmospheric Environment Service, Weather Modification Activities in Canada, January-December 1975 (Reference File 8200-1 (ARPP)).

165. Letter to Permanent Representatives of the WMO (PR 26-25) from WMO Secretariat regarding Resolution 12 (Cg-vii) (Dec. 1975).

166. Letter to Ray Jay Davis from Mason Charak, U.S. Dept. of Commerce (Jan. 1976).

167. Supra note 159.

168. Weather Modification Division, Office of Plans and Programs, NOAA,

Environmental Impact Statement, Hurricane Modification Research (June 1971).

169. *Supra* note 156.

170. *Supra* note 157.

171. Treaty on Principles Governing the Activities of States in the Exploration and Use of Outer Space Including the Moon and Other Celestial Bodies, Article 9, 61 A.J.I.L. 644 (1967). 18 U.S.T. 2410 (1967), 610 U.N.T.S. 205 (1967).

172. *Supra* note 161 at 7.

173. Lake Lanoux Case (France-Spain), 53 A.J.I.L. 156, 170, 24 I.L.R. 101 (1957).

174. K. Skubiszewski *in* Sorenson, ed., Manual of Public Int'l Law 775 (1968).

175. *See* text at note 1 *supra*.

176. 28 GAOR Supp. 21 (A/9021), 3 Report of the Committee on the Peaceful Uses of the Sea-bed and the Ocean Floor Beyond the Limits of National Jurisdiction 35 (1973). This draft article was submitted by the delegations of Ecuador, Panama, and Peru for inclusion in a convention on the law of the sea. In its present stage it, of course, binds no one. Considerable controversy surrounds the creation of a new legal regime for the world's oceans, including such questions as the status of the deep sea-bed and the proper limits of the territorial seas.

177. ICJ, Office of Public Information, U.N. Publication at 33 (June 1975).

178. Trail Smelter Case (U.S. v. Canada), 3 Rep. Int'l Arbitral Awards 1905 (1949); 35 A.J.I.L. 684 (1944).

179. *See* text at notes 76-78 *supra*.

180. *Supra* note 156.

181. I. Brownlie, Principles of Public Int'l Law 18 (2d ed., 1973).

182. MacGibbon, The Scope of Acquiescence in International Law, 1954 B.Y.I.L. 143, 147.

183. *Supra* note 163 at art. VII.

184. Barros & Johnson, The International Law of Pollution 75 (1974).

185. U.N. Yearbook 320 at 321 (1972).

The Legal Uncertainties:
A Scientist Responds

John W. Firor, Executive Director

National Center for Atmospheric Research
Boulder, Colorado 80303

Brown Morton in his introduction noted that the law deals with facts and that science is relied upon to establish the facts. Ray Davis pointed out the ironic situation of weather modifiers being protected by their own ignorance. Uncertainty prevents collection of damages and issuance of injunctions because the experts cannot show causation in fact or probable causation before the fact.

This uncertainty will lessen as we acquire additional experience. The situation raises an image of a battle between opponents who want to stop such activities or to collect damages and proponents of weather modification. At present we are hiding behind a rock—a rock labeled "ignorance"—and we must learn to run across the battlefield to another rock labeled "perfect knowledge." There is a very troublesome middle ground where we will be shot at, where partial knowledge will allow reasonable courts to stop weather modification activities or to assess damages. The difficulty with this analogy is that the second rock—the one labeled "perfect knowledge"—does not exist in the usual sense in which we use the term. We have imperfect knowledge of storms and we have imperfect knowledge of what happens when we attempt to modify them. We may learn much in the coming years, but the natural variability of weather will continue despite what we learn. Therefore, we do not and will not operate in a situation in which we can predict with great confidence on a case-by-case, storm-by-storm, mountain-by-mountain basis. We must learn to live with results that are statistical in nature. This is similar to the analogy that Lou Battan used—about taking medicine prescribed by a physician. Federal law requires that pharmaceutical companies sell efficacious pills, but efficaciousness is not determined on the entire population but rather on a small sample. We must deal with small samples in weather modification when assessing whether a particular modification attempt will be or has been efficacious. That determination will be statistical because of the high variability of natural weather patterns.

For example, assume that we are planning to augment precipitation and that we are going to do so with a procedure that was tested in a well-designed, randomized, statistical experiment with all the safeguards that researchers would like. Further, assume over a five-year period that we experienced a 30% increase in precipitation, but that on any individual day the precipitation might have been increased or decreased by 60%. Note that the 30% average increase describes only average benefits for the agricultural community. On any one farm on any one day, too great an increase could cause an individual farmer to suffer. That is what we must live with because we never will know specifically who gained and who lost from our activities.

Therefore, the evolution of law in courtrooms and in legislatures must proceed within this reality of weather modification. The facts are that both the beneficial and the harmful results will be spread statistically and that in most cases it will be impossible to say which individuals benefited and which did not.

And then we have problems concerning how scientific "proof" is reached from studying sample data and extrapolating from them to the total population. For example, in the case of DDT—as mentioned earlier by Roger Hansen—the chemical was known to be beneficial to some individuals, but the damages could be deduced only from a complex argument not involving a specific individual. We must adjust our decision-making process to the statistical nature of the evidence.

The Legal Uncertainties:
A Lawyer Responds

Howard J. Taubenfeld

School of Law
Southern Methodist University
Dallas, Texas 72575

In 1968 I organized a conference of lawyers and scientists at Southern Methodist University at which we discussed weather modification.[1] I recall that two Michigan legislators attended to learn how that state should proceed in this field. One of them took the floor during a session to comment that, since so much uncertainty surrounded the subject, Michigan should not even consider legislation on weather modification because nobody knew what to do about anything.

From what I have heard here, I conclude that conditions have not changed much. Even though we have additional experience from a number of experiments and commercial applications, no substantial scientific advances have been made since 1968.

Without question, vast uncertainties exist concerning atmospheric phenomena. In 1970 at a conference sponsored by the Massachusetts Institute of Technology,[2] I attended a session on computer modeling during which the most advanced computer model for the atmosphere was explained. It struck me as being extraordinary, although none of the physical scientists thought it amusing, that they had not, after many years and much money, advanced far enough to introduce clouds into the computer model. There were a good many reasons why that was so, but to me it seemed a little strange not to include clouds in a discussion of the atmosphere.

There are vast uncertainties that are not the scientists' fault. Natural processes are just too difficult to understand. These scientific and technical uncertainties have led to many of the legal problems that we face and have compounded the difficulty of the commercial modifier in attempting to carry out his activities. These individuals can assure customers that they will increase precipitation or suppress hail but, when they are sued, they claim that there is no proof that anybody did anything to anyone. Both comments may be accurate. As chairman of the Advisory Committee on Weather Modification of the Texas Water Development Board, I asked the technical staff some questions while considering a request by a private entrepreneur to operate generators in Texas to seed clouds with the hope of enhancing precipitation in Oklahoma. I wanted to learn if there was any evidence that we would suffer harmful effects in Texas, and the staff replied that since there were undoubtedly no effects anywhere, there could not be any harmful effects in Texas. We followed that advice, and the operator has been seeding clouds ever since. Another example of uncertainty is found in an original draft by the U.S. Air Force concerning seeding during the last drought in Texas.[3] I was amused to read that in one instance three storm cells were approaching a particular town, the outer two of which had been seeded. The town had floods and the conclusion was that the inner one, the unseeded one, had caused the flooding. I found that ability to distinguish among storm cells rather remarkable, and the conclusion did not appear in the final report.

As Ray Davis has pointed out, we see some changes in state law and, in particular, better types of regulatory arrangements. There is more interest being shown by states, and additional funding is becoming available, although South Dakota has reversed its position of being actively involved. Since 1968, there is in essence no more federal regulation. The prior reporting requirements to the National Science Foundation no longer are in effect, and modifiers now report to the National Oceanic and Atmospheric Administration. We still have no federal regulation such as licensing.

Federal research funding has fluctuated in terms of real dollars but has never been great enough to support an adequate program. One endeavor deserves notice—the hail suppression work at the National Center for Atmospheric Research. The initial reports seemed to indicate a favorable effect—in that hail was decreased—although later reports suggest that no real decrease in fact occurred and that in some instances the amount of hail may have increased, accompanied by a possible increase in precipitation. As far as the federal government is concerned as principal financer, weather modification is a "back-burner" type of operation. Financing is modest and has been relatively ineffective in producing the desired effect.

Litigation has increased somewhat since 1968 with predictable results. Due to the uncertainties of weather modification facing the persons who oppose this activity, they tend to lose in the courts when they try to stop a program or to collect damages after a modification attempt. They then turn to the political arena for satisfaction. In southeastern Colorado, where a privately financed hail suppression program was in operation, the opponents took the issue to the legislature and through the legislature obtained a revised licensing authority that has had the effect of ending commercial seeding operations.

When courts in Texas failed to issue injunctions against private operations for hail suppression and precipitation augmentation, the legislature responded with a statute that makes it more difficult for the modifier to begin operations. Public hearings are required, and it no longer is at all certain that licenses will be issued almost automatically to qualified operators. No doubt many reasons account for the decision by South Dakota to discontinue funding its vigorous state-financed program.

At the international level, Project Stormfury not only was stopped in the Atlantic for a number of reasons but also has been frustrated in its attempted move to the Pacific, where there are more storms. The Philippines appears to favor it, but Japan and China do not, and as a consequence the project has not become operational. Several participants at this conference attended a meeting recently in Washington, where, after hearing detailed descriptions of the progress being made in atmospheric modeling and in the proposals for Stormfury, we heard a representative from the Department of State say in effect, "Gentlemen, no, and no in the Atlantic because there are a lot more countries there, and if you want to do anything in the Pacific, we no longer will be satisfied with acquiescence by Japan and China—we need an affirmative assurance from each of them before we go ahead." If you think a moment about the difficulties of negotiating international agreements on complex issues of this sort, you can conclude as I have that the project seems to be dead. Project Stormfury, if it exists at all, is now contemplated off the west coast of Mexico near Acapulco, but it is quite a different program from the one planned and we have no assurances that it will work at all.

As a law professor I like the subject of hurricane suppression best. Even if the technology existed, we could think of many legal and moral dilemmas that we would face each time the question arose as to whether we should use it. Consider, if you will, a hurricane approaching the coast of Texas near Galveston. Assume also that the wind speed approaches 100 knots and that we have the technology of reducing wind speed to 85 knots by spreading the storm over an additional 15% of coastal area. Should we reduce the violent impact in Galveston at the expense of other coastal areas? Is this a good or bad action to take? How do we balance the decrease in damage to Galveston, if any, with an increase elsewhere?

It has been argued that issues of this sort justify a decision not to proceed any further with developing the technology. Even if the technology works, the problems of liability and other legal issues will be simply horrendous. I do not mean to imply that we should abandon scientific activity, for I think that weather modification can help produce necessary resources worldwide. It should be studied intensively. But I am saying that it should be studied in a scientifically rigorous and socially responsible way. This means that every research and operational program should be accompanied by a program that will provide full recompense to all parties who may suffer losses. If we do not provide compensation, the legislatures are going to succumb to pressures and stop weather modification activities—certainly the commercial ones and perhaps even the experimental ones.

There is evidence that we must expect extra-area effects. These might be beneficial but that is not inherently the case. Suppose what happens is an increase in precipitation where farmers are harvesting crops. They most certainly do not want rain. Evidence suggests that an increase in hail rather than a decrease can follow hail suppression attempts, and some damages are very difficult to assess, as conflicts in expert testimony during litigation attest. I conducted my own little study by asking scientific experts whether they thought that weather modification contributed to the floods in Yuba City and Rapid City, and I received a negative answer on the whole. But when I, lawyer-like, put the question the other way around and asked for the scientists' assurance that the cloud seeding could not have contributed to the additional precipitation and flooding, I received uniformly the answer, "No, I can't say that. They certainly could have." Without doubt, the public has a right to question the outcome of these activities.

The federal program in its broad political aspect has been mismanaged from the outset. I do not mean this in reference to particular agencies or particular projects but to general lack of coordination among agencies and to strictly political interferences. Funding and other actions by the Office of Management and Budget have been tied to political cycles and funding cycles rather than to the needs of research. Research projects in difficult subjects such as this often must run for 10 or 20 years before yielding useful results. The federal government simply is not accustomed to that sort of spending for scientific projects, although it seems to be for certain other ones.

Dr. Stever, Director of the National Science Foundation, several years ago erred rather badly, I thought, when referring to the hail suppression research at the National Center for Atmospheric Research as being the ideal RANN program.[4] He suggested that an investment of $3 million a year for five years that helps reduce hail that causes $300 million worth of damage a year is very cost effective.

A $15 million investment that returns $300 million annually is magnificent indeed, but the numbers were meaningless. The projects have been poorly designed, from my point of view, as I have been lecturing my friends in the scientific community. From the beginning, scientific projects in a field like weather modification should be designed not only by physical scientists, who do not know all the right questions to ask, but also by ecologists, economists, sociologists, lawyers, and others who will help plan the project to produce data needed to answer social and political questions as well as those related to atmospheric phenomena.

In conclusion, I think the question of whether to proceed should be answered in the affirmative. These projects should be continued, they should be enlarged, and they should be broadened in scope to provide the compensation that I just mentioned. Also, they certainly should elicit as much information as possible so that all relevant disciplines—including law—are better prepared to answer the socially important questions.

Notes

1. H. Taubenfeld, [ed.]. Weather Modification and the Law (Oceana Publ. Co., 1968).

2. W. Matthews et al. [eds.]. Man's Impact on the Climate (MIT Press, 1970).

3. Mimeographed draft for private circulation (1971).

4. Research Applied to National Needs, a funding program of the National Science Foundation.

Open Discussion

LEE LOEVINGER: I would like to comment briefly on Roger Hansen's statement that a scientific probability does not establish a fact—which I take to mean a legal fact—and I question what he means by this. A lesson of my entire legal training and experience with the rules of civil procedure, for example, is that one side prevails by establishing a preponderance of the evidence. This, to summarize quickly, means roughly 51% as opposed to 49% of the evidence. Now this, unless I am much mistaken, is closely analogous to a scientific test, perhaps like the chi-square test of significance. It seems to me that this can be reduced readily to the scientific test of probability or of statistical significance. When we speak of criminal cases, we're referring to proof beyond a reasonable doubt, but more often we deal with civil matters and a preponderance of the evidence. Law has always operated on a scale of probabilities, although verbal rather than numerical, just as science operates on a scale of probabilities. John Firor told us that the results of weather modification activities must be described in statistical terms; my understanding is that this is true of all scientific results and that no such thing as absolute certainty exists, other than statistics. Certainty is simply a hypothetical limit on the probability scale. I think our problem is to reconcile the verbal statistical scale of law with the numerical statistical scale of science.

DAVID ROSE: I agree that it is not an open-and-shut case where we have either zero effects or absolutely certain effects. One of our shortcomings is a lack of appreciation or understanding of statistical methodologies that allow us to predict results on the basis of samples, with predictable deviations around the expected result. This is confused somewhat in the case of weather modification because we seem to be dealing with an art that is even more statistically uncertain than most of the sciences.

JAMES SMITH: I would like to reinforce what has been said concerning the degree of certainty in weather modification and in meteorology in general. We should be satisfied to deal with best estimates based upon the best knowledge available. I don't see any difference between dealing with these uncertainties and with the typical uncertainties encountered in a court of law. Psychiatric testimony in criminal trials usually is conflicting. On the other hand, in the mountains of California we deal with normal variations in annual precipitation of 150% from year to year and with snowpack depths ranging from 1 to 7 meters. I would be delighted if someone could tell me how to determine the impact of a 10% increase in snowfall on a plant species that has evolved under that kind of climatic regime. Here, as in the courtroom, we must depend upon the knowledge of experts and their best judgment as drawn from available scientific evidence.

MODERATOR: I assume you would distinguish a scientist speaking as a scientist from a scientist speaking as a person who can make scientifically informed judgments, because you imply that the scientific evidence is not really very scientific at all.

JAMES SMITH: There are many bits and pieces of information that we assume are factual, and the expert is the person who can explain how the bits and pieces fit together. In many cases we understand the processes but don't know how the numbers should be assembled. Roger Hansen was wishing for the day when we had models to explain meteorological processes, but before we can formulate these models we must know the processes that drive the model.

HAROLD GREEN: A comment, if I may, on the difference between facts as viewed by scientists and facts as viewed by lawyers. The goal of science is to reach conclusions that are objectively true. A scientist, I suppose, would not be willing to accept as true something that does not have a very high degree of probability of being true. In the legal system, however, we are willing to accept as true some things that are simply more probably true than false. This is the 51%-49% balance that Lee Loevinger mentioned.

And as a matter of fact, we can go even further, because the legal system is quite prepared to accept as true some things that informed persons may regard as untrue. It is imperative that we keep this distinction concerning objectivity and truth between our two disciplines well in mind.

MODERATOR: The civil test we referred to as "preponderance of the evidence" differs from another civil test of greater certainty called "clear and convincing evidence" and the criminal test of "beyond a reasonable doubt" in that the latter two require a degree of belief by the judge or jury and not a simple mechanistic assessment of the 51%-49% balance in evidence.

HAROLD LEVENTHAL: The legal system in recent years has begun to develop standards that differ from the more rigid ones of the past. Courts by their very nature deal with evidence that is professedly debatable and with scientific hypotheses, conjectures, and speculations. As a result, courts have a variety of verbal tests, some of which require something less than a preponderance of the evidence before taking action. When the consequences of an action are very great, as in the case of carcinogenic chemicals, a lower probability of catastrophic effects will justify action by a court to protect the public. In these cases we must be satisfied with extrapolations from laboratory test animals to human populations and we do not have the epidemiological data to overcome these uncertainties. In these instances involving great potentials for harm, where the results are so horrible if an incorrect decision is made, we must act on the side of caution, and we often are satisfied with evidence of a lower order than that required in the typical civil case.

At the other end of the scale, when jurors are told that they must believe a person is guilty "beyond a reasonable doubt," they must have a moral assurance that they are correct because of the consequences of depriving a person of liberty. If the consequence of action or inaction is that the human population will be involuntarily exposed to a chemical that might be carcinogenic, the degree of evidence required certainly is much less. We deal with more than an allocation of economic harm, and courts and legislatures are becoming increasingly aware of this.

When I came to this conference I thought that it would be rather straightforward, unemotional, and noncontroversial. Now I find that it is nothing of the sort and that many questions are lurking in the wings, some of which might be very emotional indeed, as when the cattle interests and the agricultural interests differ on how to use available water. I don't know where this will lead us, but I hope it results in greater communication among members of the relevant disciplines.

As an appellate judge, I read much scientific testimony, both where scientists were speaking at ease and where they were speaking in adversary proceedings, and I am struck by the difference in vocabulary between scientists and lawyers. A scientist will introduce a significant statement with a rather modest introduction, such as, "There is some evidence to believe . . . ," as though he were advancing a very modest point. Now that clause implies something quite different to lawyers and to other persons. Along the same lines, perhaps we need a glossary to promote effective communication and to help dispel the underlying uncertainties that seem to discourage communication in the first place. The scientific uncertainties produce the legal uncertainties, with the added legal uncertainty in these cases of who has the burden of proof.

MODERATOR: Your initial statements about the probabilities required for action remind me of the analogy of a person being asked to participate in a lottery of 1,000 people, in which 999 will receive $50 and 1 will be shot. The rational person would decline, although the probability of winning is even greater than that we assign to "beyond a reasonable doubt." I am pleased that the federal courts in *Reserve Mining*[1] and *Ethyl*[2] discussed the matter of low probabilities of suffering major consequences and that the judicial responses are increasingly flexible under these circumstances.

NASH ROBERTS: After some 30 years as a consultant meteorologist, with early experience in planning weather modification operations for sugar planters and for coffee growers, and after experience in forensic meteorology as an active courtroom participant, I cannot understand how we can speak of facts as if they were associated with a probability of 100%. In reality, the effects of cloud seeding are so small in comparison to the inadvertent modification that occurs every day that it is virtually impossible to ascertain whether a weather modification program was a success. Every person who plows a field or dams a stream or lights a fire changes the atmosphere.

More nuclei are introduced into the atmosphere by an industrial plant or by a city in one day than by all the weather modifiers in the next century. This is a deliberate act that causes inadvertent modification. I just do not understand how we can evaluate the effect of silver iodide introduced from a single airplane, when we remember that the number of nuclei already present might be a function of traffic patterns in a nearby city or of shutdowns of industrial processes. Reliance on atmospheric models of the future is not satisfactory, because they will not take into consideration these fluctuations in nuclei generated by a modern society.

MODERATOR: Certainly someone would care to comment on that.

WALLACE HOWELL: I would like to mention an analogy concerning environmental effects. We found that deer mouse populations respond to depth of snowpack, but that red-backed vole populations do not. Much more important to these species is how much trapping and poisoning the U.S. Forest Service does to control the local populations of small mammals that would damage seedlings. Much more important to elk populations than depth of snow as a physical barrier to movement is whether government agencies decide to carry the herd through a period of stress by artificial feeding or decide to increase the number of elk hunting licenses. My point is that weather modification very often will cause only secondary or tertiary environmental influences and that many changes in even the directly affected environmental components are associated with activities not at all related to weather modification.

JOHN FIROR: I would like to comment on the statement by Nash Roberts, who argued that the extent of inadvertent modification is so large that it is impossible, or nearly impossible, to ascertain the effect of advertent modification. It seems that we are in the same situation as a city that applies for a permit to discharge sewage into a river already exceeding the applicable pollution standards. Two arguments are possible. We could argue that what the city wants to do is such a small change that it should be allowed, or we could argue that a river that exceeds the standards should not receive any additional discharges. We are going to face similar questions concerning the depletion of ozone in the stratosphere, perhaps even having to allocate amounts to be destroyed by supersonic aircraft, fluorocarbons, and fertilizers. We need a tremendous body of technical information to reach a political decision about how clean we want our atmosphere. How far downwind inadvertent modification extends is an example. In principle, one can say that no difference exists between the inadvertently discharged nuclei and those from controlled operations, but in the end our political decision-making process must decide how much of this clean resource should be allocated to each user.

NASH ROBERTS: I was not arguing whether we should decrease or increase weather modification activities but was simply pointing out that to want absolute predictive abilities is asking too much. It is unrealistic and, I maintain, will be unrealistic in the foreseeable future to obtain absolute guarantees that the weather modifier will produce specified results.

LOUIS BATTAN: I don't agree with the proposition that other human activities make it virtually impossible to evaluate the effects of cloud seeding. The great variability in natural precipitation—amounts and patterns—increases the difficulty of establishing whether there has been a 10% increase, for example, but it is not impossible. With increased experimental and operational data, we should be able to speak in terms of probabilities, and if reasonable people agree that the probability is sufficiently great, then we can conclude that something did happen and that it deserves recognition.

ROGER HANSEN: I have to comment on Lee Loevinger's interpretation of what I said. We have been mistakenly placed at two extremes. I said that a statistical probability would not prove an absolute physical fact. It does not prove whether something exists or does not exist. I do believe, however, that neither law nor science insists upon proof at the 100% level. Obviously, it is ridiculous to attempt to adhere to an absolute standard of proof, and I began my remarks by recognizing that our well of ignorance is deep. I would not possibly suggest that we had to meet an unobtainable standard of absolute certainty.

LEE LOEVINGER: Quite frankly, I still don't know what Roger is talking about. If there is such a thing as a fact that is independent of either scientific or legal proof, it is in the realm of metaphysics or spiritualism and has no place in this discussion. We are talking about legal and scientific standards of evidence, and I repeat my belief that the basic problem is how to reconcile the probability scales of law and science. If Roger wants to believe that there is a metaphysical fact existing out there that is, in a Kantian sense, unrelated to these scales of probability, he certainly is at liberty to do so, just as he is at liberty to hold any religious faith he desires, but I don't think that it is relevant to our discussion here.

ROGER HANSEN: I don't want to get into an adversary proceeding. That would be aimless, but it is apparent that we also have a communications problem between lawyers and lawyers.

NASH ROBERTS: To summarize my position, if anyone can give us a definite probability—verbal or numerical—that a weather modifier induced rain or suppressed hail, I would consider it a triumph in our field.

EDWARD COLLINS: I have two points to make on this general issue. First, we must consider the forum in which any discussion concerning probabilities takes place. In the civil case, the judge or jury is asked to decide on the preponderance of evidence presented by the parties. But these issues frequently are discussed in a public forum with no controls over the quality of evidence or its relevance. In the public forum, the uncertainties are very large indeed and in many instances we cannot even hope for quantitative evidence, so resolution of any issue discussed in that forum will probably not be aided by mathematical probabilities.

Second, we must keep in mind the burden of proof. Roger Hansen this morning mentioned the *Reserve Mining*[1] case, in which the company polluting the atmosphere and Lake Superior with a suspected carcinogen argued that it should be allowed to continue until there was a preponderance of evidence that the pollutant was causing cancer. But suppose the rule was that the company must establish by a preponderance of evidence that it would not cause cancer? The issue is not entirely the standard of proof that we require, but rather it is a question of who has the burden of meeting it.

ARNETT DENNIS: I have just witnessed South Dakota's decision not to fund weather modification programs this year, and the processes over the past several months through which this was accomplished—through citizens' organizations, legislative hearings, letters to the editor, and so forth. I came to this meeting hoping to be enlightened about the public decision-making processes that are going to be involved in future years in this controversial area, but I am ready to conclude that our courts and legislatures make decisions in many other controversial areas without the degree of certainty we might like. The advantages and disadvantages of a new shopping mall, an urban renewal program, a new freeway interchange, and many other issues are decided in a public arena that is messy at times but yet yields a decision. These decisions are not always satisfactory to everyone, but in our society we learn to go along with them.

Would some of the lawyers here care to comment on any lessons we can draw from other complex areas and predict what we might see as a decision-making process in weather modification activities over the next five or ten years?

RAY DAVIS: The law deals with exceedingly complicated issues on a regular and routine basis. It has been asserted that the problems of causation in weather modification are so difficult they should not be entrusted to an ordinary jury. It may be that these problems of causation are very troublesome, but we regularly rely on the decisions of jurors in cases that are just as complicated as some of the ones we are talking about here. A motor vehicle accident, to name a common one, often raises complicated issues of causation. Our governmental processes handle questions of this complexity all the time, and we are not unaccustomed to making decisions in the middle of the probability spectrum. No reason exists why we should not be able to do so with weather modification in the next five or ten years.

BARBARA FARHAR: The public is not accustomed to having anyone make decisions for them concerning the weather, as it is for public works and similar projects. The public is not familiar with the consequences of the decisions and does not have the necessary experience or confidence in the record of decision making within any existing institutional arrangement in the case of weather modification. The public is not familiar with the decision-making mechanism for this type of far-ranging activity, and this lack of familiarity means that the decision-making process is not routine. With time and experience, the public may come to accept our traditional methods with regard to weather modification, or more innovative decision-making mechanisms might be devised.

HOWARD TAUBENFELD: Before we talk about what the law is going to be doing with weather modification ten years from now, we first must get an answer to the question that I have been putting to my scientific colleagues for some time: "What is the science or technology of weather modification going to be doing ten years from now?" Since meteorologists are extraordinarily reluctant to forecast beyond the next three days, I don't think we will get a very satisfactory answer to that question. It is unreasonable to ask lawyers to do too much, particularly since it appears that the scientific answers will not be forthcoming any faster than they have been over the past ten years. I predict that lawsuits will continue to fail, on the whole, and that legislatures will become the forum for legal decision making in this area. A number of states will bar weather modification activities or make them so difficult that they cannot be carried out effectively. I further predict that we will see an increase in the federal role in all aspects of weather modification. Now, those forecasts are much more specific than any that my scientific colleagues have been willing to make.

HAROLD LEVENTHAL: We now are discussing just what the law does. It expresses the values of society. We all realize that it is really a combination of values and facts, but when the facts are unknown or unknowable, we must resort to a consideration of our fundamental values.

Tradition plays a part, as in the judicial acceptance of a laissez faire philosophy that has been generally perceived to be for the public good. In these traditional areas we will encounter a momentum in favor of continuation. If, on the other hand, we are doing something different or proposing to do something different, we encounter moral values. Many businessmen have a motto, "If you don't know what to do, don't do it." If they adhered to it, there would be no business decisions ever made. When I was in private practice, I was always irritated by clients who wanted me to predict the law with a degree of certainty far greater than the certainty they would demand about their advertising budget or their sales programs. They made business decisions in an atmosphere of very great uncertainty indeed, yet they were dissatisfied to learn that many features of the law were so uncertain. It is evident to me that if science and technology cannot resolve the uncertainties concerning weather modification through sound experimental programs and through convincing operational achievements, then the law will react in a way that protects the values of society. The question then arises, "Is there any reason why that is not good?" If people are suspicious about weather modification and their suspicions cannot be put to rest, then the traditional values will govern our social decisions. This will result in slower implementation of knowledge, and in fact slower accumulation of knowledge, but is this always a bad result from society's point of view? Is it so bad that we slow down to let knowledge catch up? Is it not characteristic of the scientific method to proceed rather slowly through a series of successive approximations?

RUTH DAVIS: Two points seem to be emerging from this discussion that are not within the domain of either science or law. One of these is that our traditional way of determining values in the United States is to see what happens in private enterprise. We should be observing carefully the extent to which the services of weather modifiers are in demand and if weather modification is becoming a marketplace service or commodity. We have not discussed whether the number of commercial operators is increasing or decreasing. I gathered from informal conversations that it is not increasing and that we do not have any sense yet that they will be recognized within the free enterprise system as having great utility.

The second issue is much more important. We apparently are confusing what we want scientists to guarantee with what we should be asking technologists to guarantee. We have been speaking of scientific proof when we really should be asking for guarantees of technological application. If we expect science to provide guarantees or warranties before the experiments have been conducted, we are defeating the purpose of science. We should be paying more attention to the interactions among scientists, technologists, lawyers, and businessmen. We are mixing the science of atmospheric physics with the technology of weather modification.

DANIEL BRONSTEIN: I think we may have overexpanded our discussion and have been talking too much about public policy questions rather than law and science interactions. We are discussing science as a determinant of public policy that then influences law. Nobody makes political decisions rationally, and all surveys and other investigations reinforce this. The most important decisions we make in life are made emotionally, and this carries over into the public decision-making process. Due to this emotional basis for action, I don't know if we can analyze satisfactorily the relationships among law, science, and policy. If we want to discuss science and law, I think we will have to retreat a bit from the policy area.

LOUIS BATTAN: I would like to comment further on what Judge Leventhal said about courts serving as agents in determining social values and about how this process might show that society wants scientists to slow down for a while, or at least go back to the laboratory. Meteorology is one of those sciences that cannot be understood from laboratory research alone. We have to work in the atmosphere, and the future of this science cannot be determined in a courtroom either. If certain lawsuits result in the wrong conclusion and, as a result, put an end to reasonable experiments in the atmosphere, we will be unable to establish the scientific facts or scientific probabilities.

MODERATOR: It is not an enviable position to be the allocator of time at a conference like this. We thought it would be helpful at this stage to discuss specific applications of weather modification, with emphasis on the scientific and legal uncertainties and on ways to resolve them. John Firor now will present a paper prepared by David Atlas, his associate at the National Center for Atmospheric Research, on a hail suppression project. This evening, Dean Mann will describe the Yuba City episode that resulted in extended litigation.

Notes

1. Reserve Mining Co. v. Environmental Protection Agency, 514 F.2d 492 (8th Cir. 1975).
2. Ethyl Corp. v. Environmental Protection Agency, 541 F.2d 1 (D.C. Cir. 1976).

Hail Suppression: Uncertainties, Risks, and Their Implications[†]

David Atlas*

National Center for Atmospheric Research
Boulder, Colorado 80303

Introduction

I was asked to present to this conference an example of a large-scale weather modification activity that is technologically feasible and that would affect more than one state. Although not all phases of the hail suppression project I shall describe are technologically feasible at present, the doubts in this regard should make it even more pertinent to the objectives of this conference. In many respects, the National Hail Research Experiment (NHRE) might be considered the model of an ultimate operational hail suppression program. However, most commercial operators would consider it excessively sophisticated and thus too costly for application. This may be true, but when you understand the uncertainties surrounding hail suppression activities, you might regard our elaborate facilities as necessary safeguards for conducting a nonhazardous operation and for gathering sufficient information in the event a lawsuit results. I shall describe a hypothetical hail suppression program patterned after some of those now in operation, with added features based on the NHRE experiences. I then shall discuss the key uncertainties, the potential hazards involved, and their possible social, political, and legal consequences.

Basic Types of Hailstorms and Hail Growth Processes

The first storm type is the single cell or multicell ordinary hailstorm, which is characterized by an evolution from (a) young new cumulus turrets to (b) radar-detectable cells aloft that contain the first signs of precipitation particles to (c) the mature cell in which hail grows to its maximum size and falls and radar reflectivity reaches its maximum intensity to (d) the decaying stage in

---

[†]John W. Firor presented this paper in the author's absence.
*Current address: Goddard Space Flight Center, Greenbelt Road, Greenbelt MD 20771

which updrafts have been largely replaced by downdrafts and precipitation intensity is reduced. Each stage usually lasts about 10 to 15 minutes, so the lifetime of a typical storm is 40 to 60 minutes. However, continuous development of new cells within a storm system may make the storm last considerably longer. At any one time, the storm complex may be comprised of a multitude of such cells in various stages of evolution.

Most of the hailstones appear to remain within a single cell as it evolves. The initial growth of cloud particles occurs by a combination of condensation on appropriate nuclei and subsequent coalescence with other cloud droplets. In warm-base clouds with modest updrafts these may grow to supercooled raindrops that subsequently freeze on freezing nuclei and become the embryos of hailstones. In cold-base clouds, the larger cloud droplets may freeze or ice crystals may form; in either case these grow to snow pellets by collection of other supercooled cloud drops and also become the embryos of hailstones. Thus, the embryos of hailstones may be either frozen raindrops or snow pellets depending upon the cloud conditions and temperatures. Further growth occurs by collection of supercooled water carried by the updraft, and final growth occurs while they are nearly in balance with the updraft near the top of their trajectory. The maximum stone sizes are very close to those that can be supported by the draft, and little further growth occurs as they fall to the ground because of the absence of a supply of cloud water. Rather, some melting is likely.

The most difficult stage of particle growth to explain is that between the micron and the millimeter size particles. We believe that this growth can occur naturally in a 10- to 15-minute period only if large "favored" aerosol particles (*e.g.,* dust) of about 50- to 60-micron diameter enter the cloud base within the weaker portions of the updraft.

The second storm type is the quasi-steady-state supercell storm. One studied by Browning and Foote[1] lasted more than 9 hours, traveled over 240 miles (390 km), and resulted in severe hail damage in a path almost 180 miles (390 km) long by 9 miles (15 km) wide. Hailstones over 3 inches (8 cm) in diameter were common. This type of storm is characterized by an exceedingly strong, steady, and tilted updraft that enters the so-called weak echo region or vault. Although the weak echo vault marks the region of maximum updrafts and maximum supply of supercooled water, the cloud particles there are carried up so rapidly that growth to precipitation and radar-detectable sizes is precluded; thus, the absence of radar echo. Most of the particles carried up within the strong updrafts in the vault become small ice crystals at the higher, colder levels and are evacuated with the strong high-level winds to form an anvil cloud that may extend many miles out ahead of the main cell. However, particles that rise on the weak flank of the updraft do have time to grow. Some of these fall out on the forward side of the vault to form a forward overhanging region called the "embryo curtain." These larger falling particles or embryos may be caught up once more in the tilted updraft, this time to be carried over the vault where they are in approximate balance with the updraft and where they receive an abundant supply of supercooled water and grow to large hailstones. These generally fall to the ground on the left rear side of the vault.

Considering the gross differences in storm structure and hail formation in the two storm types, we assume that the methods of seeding must differ and that the efficacy of seeding must be evaluated differently.

The statistics on the frequency of occurrence of the basic storm types are sketchy, but typically only 7% to 11% of the storm days account for 50% to 75% of the seasonal hail fall and crop damage in the U.S. High Plains. In northeastern Colorado, we may expect about half of the total season's hail to occur on an average of 1 to 2 days. One storm day accounted for 70% of the total hail damage during a hail suppression project in the northern Caucasus of the U.S.S.R.[2] Similar frequencies prevail in most hail-prone regions of the world. While most storms consist of evolving ordinary cells, the most damaging ones in Colorado, Oklahoma, and Alberta often are supercells.

The significance of the preceding is fourfold: (a) The great natural variability in hail fall requires a very large sample of both seeded and unseeded storms to detect with confidence any differences in damage done by them; this might require 10 to 15 years of data collection; (b) since a few storms produce most of the damage, an operational project must suppress the hail in these to be successful; (c) because it is difficult to control supercell storms—indeed, seeding may increase hail in these storms[1]—one must handle them with caution and perhaps avoid seeding altogether; and (d) the natural variability in hail fall makes it exceedingly difficult for a person who suffers hail damage to prove legal causation due to seeding; the modifier easily can show cases of greater damage caused by untreated storms.

Concepts of Hail Suppression

Hail storms grow from a few "favored" embryos. There generally are abundant natural embryos present, but two kinds of natural selection mechanisms produce a favored few. (a) In ordinary cells, the rapid growth of some embryos causes them to be the first in time to encounter undepleted cloud water; and (b) in supercells, the recycling embryos that penetrate farthest into the updraft core are the first in space to encounter undepleted cloud water.[3]

Thus, to suppress hail we must preempt the development of the "favored few." The preferred way is to promote "beneficial competition" by artifically introducing more embryos to compete for the available water with the result that none grows large.

Another seeding approach, known as "glaciation," is to seed the super-cooled water supply of the cloud so that the resulting ice crystals bounce off the larger pellets and impede their growth in this manner. However, it generally is conceded that suppression by glaciation requires economically and logistically infeasible amounts of nucleating material. In fact, seeding by bene-ficial competition is accompanied by some degree of glaciation that, depending upon its magnitude and location, may be either beneficial or detri-mental. It is now generally accepted that seeding should be done in the younger and weaker updraft stages of ordinary cell storms and in the weaker updraft fringes of the steady-state supercells. But there are two hazards in this approach, particularly in the case of the supercell. Inadequate seeding rates or poor targeting of the seeding locale may produce more rather than fewer embryos that reach the edge of the weak echo vault where they can also grow to large hail in the abundant water supply and thus produce more rather than less damaging hail.[1] On the other hand, excessive seeding rates and inadequate targeting may cause excessive competition for the water supply so that few particles grow sufficiently to fall out; rather they may be carried up and evacuated in the anvil thus reducing much needed rainfall at the ground.

Another problem with hailstorms in the High Plains is caused by relatively cold cloud bases and short distances of fall to the ground. These conditions ensure that most hailstone embryos will be snow pellets rather than super-cooled raindrops; but present seeding methods are much more effective if the competitive embryos are cold raindrops rather than snow pellets. Moreover, work with simulation models suggests that hail is decreased by seeding only in the warm clouds, that it is increased by seeding cold clouds, and that greater seeding rates yield more hail.[4]

Thus, while we have learned a great deal about the nature of hailstorms, many uncertainties remain. Nevertheless, the possible hazards noted above also imply legal questions as to causation and the use of all due caution in con-ducting suppression operations in spite of our still rudimentary knowledge.

Hail Suppression Methods and Results

Most hail suppression methods used by experimental and operational pro-grams employ the beneficial competition concept, although some programs have tried glaciation. As noted earlier, the two are not clearly separable. The methods differ in the means of delivering the seeding agent, the amounts delivered and, to a lesser extent, the type of agent and the means of targeting the seeding zone.

Most programs use radar to detect incipient hailstorms, but the reliability and manner of use varies greatly. In some operational programs, radar is used only as a qualitative guide, with the choice of which clouds to seed being left almost entirely to the pilots. Information from single- and dual-wavelength radar is analyzed with elaborate formulas in the U.S.S.R. to determine the probability of hail and the rate at which rockets should be fired to deliver the seeding agent. Emphasis is placed in the NHRE upon real-time computer-generated three-dimensional radar reflectivity patterns for declaring a "hail day," identifying storm structure, selecting optimum seeding locations, and vectoring aircraft.

The size of the "protected areas" ranges from a minimum of about 23 square miles (60 km^2) in South Africa to about 18,500 square miles (47,900 km^2) in Alberta. In the U.S.S.R., a typical protected area is 1,000 to 2,000 square miles (2,600 to 5,200 km^2). The NHRE target area is 625 square miles (1,600 km^2).

It is difficult to obtain unqualified results from any hail suppression program, either experimental or operational. Reports from the Soviet Union claim 70-80% damage suppression and 4:1 benefit/cost ratios,[2,5,6] but these generally are not accepted outside the U.S.S.R. because of inadequacies in their evaluation schemes. Such inconsistencies characterize the word-of-mouth reports from visitors to the U.S.S.R.

A recent report[7] based upon on-top seeding by jet aircraft in South Africa suggests a 22% reduction in the ratio of tobacco crop destroyed to the total area hit, but this may be questioned because it is based in part on use of historical records that might not accurately reflect the annual variations in hail frequency.

A recent survey of five operational projects in the U.S. and South Africa reports reductions in crop losses ranging from 20 to 48%,[8] but these are also controversial[9] because only one of the projects included any degree of randomization, all the statistical tests were selected after the fact and a variety of tests were performed, and most of the results are based upon crop hail insurance records, the validity of which is in doubt because of the typically small fraction of insured farms. In contrast, indications of increased hail occurrence on seeded days has been found in Switzerland[8,10] and under certain meteorological conditions in Argentina.[11] Also, in three years of experimental seeding by NHRE, an average of 60% more hail occurred on seeded days.[12] However, the small storm sample also renders this result questionable due to the high variability of natural hail falls and the difficulty of predicting what would have occurred had the seeded storms not been treated.

It is also possible that the apparently discouraging results experienced by NHRE reflect the dominance of supercell and cold-base storms, which may be susceptible to increased hail when seeded, or that the seeding methods, which generally targeted the most intense updrafts at cloud base rather than the feeder clouds of multicell storms or the weak updraft fringes of the larger storms, were inappropriate.

The overall assessment of the efficacy of hail suppression is uncertain. Despite criticisms of Soviet methodology, they still provide the most persuasive support for the belief in the efficacy of hail suppression. A five-year trial of the Soviet methods now is being undertaken in Switzerland to verify their claims and to test the transferability of those methods.[5] Preliminary results from South Africa also are encouraging, as are those from some other projects, but other results are not. Moreover, there is growing concern that certain kinds of storms may be susceptible to increased hail as a result of seeding—e.g., the supercell. Thus, whatever one concludes about the efficacy of a particular seeding operation, it is not at all evident that those conclusions are valid for other regions, storm types, or methods.

A Hypothetical Operational Hail Suppression Program

To illustrate the nature of a large-scale operational hail suppression program and the problems associated with it, consider the 35-county hail-prone area overlapping portions of northeastern Colorado, southwestern Nebraska, and northwestern Kansas. The incidence of hail and the agricultural economy of this area have been studied by Borland.[13] While cropland represents about 47% of the total hypothetical operation area (HOA), only about one-half (23%) of this is planted each year. As a first approximation to a formula for allocating costs among the three states, we might use the percentage of total harvested land, but neither the hail threat nor the crop values are uniformly distributed through the HOA. A more complex formula based upon actuarial statistics of damage, analogous to the present method of computing hail insurance premiums, undoubtedly would be fairer.

Insurance records for 1957 to 1974 show that an average of 1,060,000 acres (429,300 ha) are insured annually, or not quite 17% of the average planted area. The average annual insured loss is $3,820,000, so the annual loss would be about $23,000,000 if all harvested acres were insured. Because the average farmer insures for only 32% of full value, the estimated annual crop hail losses in the HOA average about $72,000,000 or about 9% of the estimated total crop value of $834,100,000. Thus the estimated average annual loss is $11.30 for each harvested acre, compared to an average crop value of $131 an acre ($27.91/ha and $342/ha, respectively).

With total operational costs for a hail suppression program estimated between $5,000,000 and $7,000,000, the cost for each planted acre is between $0.78 and $1.09 ($1.93 to $2.69/ha) or 6.9% to 9.7% of the average loss, respectively. In other words, hail damage would have to be reduced between 6.9% to 9.7% to break even. This is a modest goal and represents a strong stimulus for farmers to participate in the program. The potential benefit/cost ratios for hail damage reductions of between 20 and 60% range from about 3 to 9, respectively, at the lower project cost and from about 2 to 6 at the higher one.

While we have increased the cost of the operation to more than twice the $0.30 per planted acre estimated by Borland and Snyder,[14] the required reduction in hail to break even is slightly less than they estimated because the average loss per acre is now 3 to 4 times the value they used. The benefits might be increased if the hail suppression were accompanied by rainfall augmentation or reduced if rainfall should be decreased. Borland and Snyder showed that a 5% decrease in rainfall approximates the economic effect of a 20% decrease in hail damage.

The following outline of the hypothetical project is described to establish a context for the political, legal, and social issues that might be raised by a large-scale hail suppression program. Other configurations might be more suitable in practice.

1. The operational headquarters would be located at an airport in northeastern Colorado that could accommodate small jet aircraft and would include a complete meteorological and forecasting station with a radiosonde, a high-power meteorological radar, an aircraft surveillance and control radar, a computer, a satellite receiving station, and the communications center. The radars would cover a radius of 150 mi (240 km) and include 35 counties in southwestern Nebraska, northwestern Kansas, and northeastern Colorado or an area of 42,850 square miles (111,000 km^2). Full operations would be conducted for the 3½-month hail season from May 15 to August 30.

Meteorological data from outlying stations would be telemetered to the headquarters for incorporation into the national data set. Radiosonde data would be computerized for numerical predictions of the probability of hail, the intensity of the storms, and the maximum likely hail sizes. Forecasts would be made of the probable times and areas of storm activity. Based upon these predictions, the operations director would declare a "go," "'stand-by," or "no-go" day and select one of a preselected number of operational plans. This information would be communicated to all substations by closed-circuit TV, with which each substation would report its operational readiness. About 50 automatic telemetering weather-observing stations would be deployed throughout the area, providing a station density of about one for each 860 square miles (2,200 km^2).

2. Five substations would be spaced more or less evenly in the HOA at airports suited for small jet operations. Each would be equipped with a medium-power meteorological radar, a one-operator radiosonde, and standard meteorological equipment. An aircraft transponder-interrogation radar would be ancillary to each meteorological radar so that aircraft position could be displayed at both local and operational headquarters. All aircraft vectoring and seeding operations would be controlled by the local operations director for his

sector, but the headquarters director would be responsible for coordination of operations from one sector to another. Local control of seeding operations is necessary not only to assure better radar observations of storm structures but also to permit simultaneous operations when necessary in all sectors.

3. Aircraft would include six turbo-charged conventional aircraft equipped with AgI-acetone burners, one based at each substation and at headquarters, and six Learjets equipped for on-top seeding with droppable flares. The jets could be based at the headquarters airport until strong convective cloud activity was reported in any of the sectors, when they could be dispatched to the substation nearest the incipient storm for immediate flight upon detection of the first radar echoes. Alternatively, if the forecast called for a front to pass progressively across the HOA, all aircraft could be deployed accordingly and also move progressively across the area.

In any operation, one aircraft would be used for visual reconnaissance to assist the local radar controller in identifying regions of new growth and to interpret visual cues not apparent on radar. The jet aircraft would be used for on-top seeding until new cloud turrets no longer were discernible from above. At that time, the other aircraft would continue seeding at cloud base. Because of the unlikelihood of simultaneous storm activity over the entire HOA and the quick response time of the jets, it is reasonable to expect that the entire area could be protected by only six jets and six conventional aircraft.

4. Because planners of any operational project should demonstrate the efficacy of their efforts and because some ancillary research will be required to optimize the methods as the program proceeds, one area would be set aside for randomized experiments. Two counties in southwestern Nebraska with a total area of 2,080 square miles (5,400 km^2) would be suitable for this purpose. Perhaps one-third of the hail days would be left unseeded on a random basis as natural controls against which the seeded days could be compared statistically. This experimental data would be heavily instrumented. Study of these storms would improve our understanding of the hail growth process and the efficacy of seeding methods. Studies of the relationships between hail size and crop damage would also be conducted here, as would investigations of the ecological effects of seeding materials on crops and animals.

If farmers in the randomized test area believed hail suppression to be effective, they might want to seed all the storms. They could be given some incentive, perhaps in the form of reduced payments, to allow some storms to remain untreated. Conducting some controlled experiments to permit proper evaluation of the overall project is so important that it is an absolute requirement of the program.

5. The cost of the entire program is difficult to estimate. Based upon 1972 costs of several existing programs and estimates of the economies of scale,

Borland[13] estimated an annual cost of $2,250,000. This is about $0.09 an acre overall and about $0.34 a planted acre ($0.22/ha and $0.84/ha, respectively). However, in my opinion, most of the projects used as a basis for that estimate were marginal. The one operational program that comes closest to the anticipated needs is that of the Colorado International Corp. in South Africa.[15] It includes a high-quality meteorological radar, an aircraft interrogation and control facility, a radiosonde station, a minicomputer, two Learjets, and seeding materials, and is sufficient to cover an area of about 5,000 square miles (13,000 km^2) with moderately frequent storm activity. The cost of this operation is about $600,000 for a 4-month season. It would not be necessary to increase the number of jet aircraft in proportion to the size of the HOA, but the additional facilities and staff required for an efficient operation and for work in the experimental area would bring the total to between $5,000,000 and $7,000,000 annually.

Political, Legal, and Ethical Issues

A host of political, legal, and ethical issues surrounds such a large-scale interstate hail suppression program. These would be exacerbated if the project also were aimed at rainfall augmentation or if there were any evidence that the hail suppression efforts depleted the already marginal rainfall. Most of the political and legal issues have been considered at length by others,[16, 17, 18] so I shall restrict my discussion to a few of the key questions.

1. State Versus Federal Enabling and Regulatory Legislation

It makes good sense to deal with weather management as an interstate problem that can be regulated best on a national level.[18] The tri-state HOA serves to emphasize the importance of federal regulation. While it is possible to conceive of an interstate compact that would unify presently disparate weather modification statutes of the three states, interstate conflicts are bound to arise about (a) the distribution of costs, (b) the seeding of storms in one state that might deprive an adjacent state of water, and (c) the responsibility for damages attributed to seeding.

Moreover, it seems logical to delegate to a single agency the responsibility of minimizing the hazards or potential side effects; establishing uniform licensing procedures, standards of performance, and reporting requirements; and avoiding conflicts among state weather modification programs. Because of the sensitive nature and possible risks of weather modification activities, I also urge that operators be certified by a body capable of evaluating their knowledge of the field. For example, appropriate legislation might require certification by the American Meteorological Society.

Until the efficacy of hail suppression without loss of rainfall is established, a test of efficacy should be required of every operational program. Because of the obvious motivation of the operator to produce evidence of net benefits, such evaluations should be conducted by an independent body. This also suggests a federal regulatory agency with competent resources to gather and analyze data, set standards of performance, and define what constitutes proof of efficacy and of the absence of deleterious effects.

2. Public Participation in Decisions and Financing

Through public surveys, Haas found that local citizens desire a strong voice in the decision to proceed with any weather modification project but that they realistically expect the decision to be made either at the state or federal level.[19] It seems reasonable that those who stand to benefit from the program should contribute to its cost. Perhaps some form of election or petition on a county-by-county basis could be required, similar to the procedure in North Dakota.[18] However, since it would be virtually impossible to prevent some effects in nonparticipating counties, the decision could be made on the basis of either a simple or a two-thirds majority of the counties in the modification district. This decision then would require participation by all counties.

The manner of financing the program is another complex and sensitive matter. Where crops are homogeneous and exposed to the same threat of damage, each farm could contribute in proportion either to the area of crop-land or to the previous year's total crop value, as done in a South African tobacco cooperative.[15] However, over an area as large as the HOA, neither the crop value nor the hail threat is uniform. Thus, a more complex payment formula based on actuarial statistics of average annual losses during the pre-ceeding decade probably would be required.

3. Injunctions and Liability for Damages

Perhaps the most critical legal problems arising from such an operational program relate to real or perceived damages due either to increased hail or to decreased rain. While experts have noted possible increases in hail and decreases in rain as a result of seeding, only recently have some authors explicitly enunciated the probability of increasing hail by seeding either super-cell storms or certain types of cold-base storms.[4, 9]

My view is in accord with that of some other authorities in the field, of which the following is representative:

> I am of the opinion that [the numerical simulations] support the possibility for hail modification by changing the available nuclei. However, whether the change increases or decreases the hail size, number, duration, etc., depends upon all components of the storm, including the pre-existing nuclei.[20]

Increased acknowledgment by experts that seeding might increase damage raises the serious possibility that hail suppression—and rain augmentation too—might be found by courts to be "ultra-hazardous" because of the extra-ordinary nature of the activity and the great damage that might result. This would overcome the necessity of proving negligence by the operator.

On the other hand, if the seeding of convective storms were not found to be ultra-hazardous, the plaintiff would have to prove both causation and negligence. The question before the court would be, "Did the operator exert due care in the light of the circumstances and knowledge available to him?"

With respect to both causation and negligence, the increased body of scientific knowledge about hailstorms and of the effects of seeding places us on the horns of a dilemma. On the one hand, the knowledge that certain storm types or seeding methods may produce increased damage is a notable advance that provides the basis for new approaches and solutions. On the other hand, this knowledge provides a stronger foundation for legal arguments of causation and negligence. Unfortunately, these problems apply to both experimental and operational programs, although they are of graver concern in the latter.

The problems of liability make it necessary to consider some form of insurance for damage associated with seeded storms. But insurance might be prohibitively expensive if there were no means of bounding the claims, either in terms of area or time of occurrence. For example, what is to prevent a farmer 100 miles (259 km) downwind of a seeded storm from claiming that the storm damaged his crops several hours later? To assist in bounding such claims, the seeding material might be tagged by an easily identifiable tracer. The presence of the tracer in a hail sample collected by the farmer would provide some evidence about whether the damage resulted from a seeded storm. The claim then might be paid, even though it could not be demon-strated that the damage exceeded that which would have occurred naturally, or the absence of the tracer could be persuasive evidence that the damage could not have been caused by the seeded storm.

4. Ethical Issues

Implicit in the above are a variety of ethical questions. I pose two of them and present my own views in response.

First, in view of the probability of increasing damage and decreasing rain-fall under some conditions, should we continue to experiment with hail sup-pression until we can provide greater assurance of minimizing such harm, particularly to the economically marginal farmers?

Because I believe that weather modification ultimately will be to society's benefit, I favor the continuation of hail suppression and rain augmentation experiments. I also see little chance that operational programs will be sus-pended or stopped. However, it is not fair to ask individual farmers in the experimental area to assume the burden of risk in order to advance science and technology for the general good. Either the farmers must participate voluntarily in the informed hope that the benefits will outweigh the costs or some form of liability insurance should be provided by the sponsoring agency. This probably means that the federal government will have to provide insurance in those areas where the risks are thought to be significant.

Second, what are the responsibilities of the commercial weather modifier in selling and performing hail suppression activities where risks are involved? While it is undoubtedly too idealistic to expect, commercial operators should provide potential customers with a realistic assessment of both the potential benefits and risks. In this regard, a federal regulatory agency might have to limit the kinds of advertising that operators may use to assure that the public is adequately protected. If operators are convinced that the benefits outweigh the costs, they should be willing to offer insurance coverage as part of the hail suppression program. Indeed, at this stage of our knowledge, and perhaps ultimately as well, it would seem desirable to combine hail suppression and insurance in a single package. This combination, as available in one cooperative program in South Africa, is backed by a reinsurance program through Lloyd's of London in the event that an unusually intense hail season exhausts the cooperative's insurance pool. If operators believe their services are beneficial, they should also be willing to incorporate tests of efficacy in their programs. Certainly their credibility would be greatly enhanced by doing so. Government regulation probably is required both to prescribe the nature of the tests and to conduct an independent evaluation.

Summary and Conclusion

The described hypothetical hail suppression program would cover an interstate area of 43,000 square miles (111,400 km^2) and would employ the most efficacious concepts and methods. The program would use modern methods to minimize hazards through the identification of storm types that should not be seeded and the development of modified seeding techniques. An area of modest size would be set aside for randomized experimentation.

The estimated cost of the program is $5,000,000 to $7,000,000, or $0.78 to $1.09 for each planted acre ($1.93 to $2.69/ha) in the area, compared to an average annual hail crop loss of $72,000,000, or $11.30 for each planted acre ($27.90/ha) at 1974 prices. Thus, only 7% to 10% of the hail damage needs to be suppressed on average to break even. A 30% suppression would result in a benefit/cost ratio between 3.1 and 4.4, depending on the above estimated cost range and assuming that no changes in rainfall resulted from the seeding. Increases in rainfall during the growing season would enhance the benefit, but a 5% reduction in rainfall might negate the benefits of a 20% reduction in hail.

Serious concerns have been expressed that seeding certain kinds of storms or using inappropriate methods may increase hail and decrease rain. This is particularly true of supercell storms, which occur perhaps only once or twice a season but may cause 50% of the seasonal crop damage. Cold-base storms may also be susceptible to increased hail through seeding. While encouraging suppression results apparently have been attained in the Soviet Union and South Africa, there is reason to doubt that similar results can be attained in the High Plains, where the seasonal hail fall is dominated by supercells and cold-base storms. In short, neither the methods used nor the results reported in the U.S.S.R. are necessarily transferable to other climatic regions.

The tri-state hypothetical operational program raises a variety of political, legal, and ethical issues. Federal legislation appears necessary to unify licensing and reporting procedures, standards of performance, and fair trade and advertising practices, and to provide independent evaluations of efficacy and potential side effects. The degree of public participation in all decisions related to modification attempts needs to be resolved.

The most sensitive issues are legal and ethical questions arising from the scientific uncertainties, especially those about possible increases in hail and decreases in rainfall. These concerns appear to increase the chance that courts will find that seeding to suppress hail is an "ultra-hazardous" activity. Alternatively, courts might establish a workable basis for proof of causation and of negligence.

The ethical questions involve subjecting farmers and other property owners to possible risks in either experimental or operational programs. While hail suppression experiments should continue in order to attain the ultimate benefits they promise and to overcome any associated hazards, ways are required to insure the public for damages. In experimental projects, this may mean federal insurance. In operational programs, the payments for hail suppression and for damage insurance could be combined.

Acknowledgments

This work was performed in part under the National Hail Research Experiment while I was its Director and was sponsored by the Weather Modification Program, Research Applications Directorate, National Science Foundation. I am indebted to John J. Snyder at NCAR for providing current economic and hail damage data.

Notes

1. Browning, K. A., & G. B. Foote. 1976. Airflow and hail growth in supercell storms and some implications for hail suppression. Quart. J. Roy. Meteor. Soc. 102:499-533.

2. Marwitz, J. D. 1973. Hailstorms and hail suppression techniques in the U.S.S.R.—1972. Bull. Am. Meteor. Soc. 54:317-25.

3. Browning, K. A. 1977. The structure and mechanisms of hailstorms. In G. B. Foote & C. A. Knight [eds.]. Hail: A review of hail science and hail suppression. Meteor. Monogr. (in press).

4. Nelson, L. D. 1976. Numerical stimulation of natural and seeded hail-bearing clouds, p. 371-77. In Proc. 2d WMO Scientific Conf. on Weather Modification. Boulder, Colo., Aug. 2-6, 1976.

5. Federer, B. 1977. Methods and results of hail suppression in the European area and in the U.S.S.R. In G. B. Foote & C. A. Knight [eds.]. Hail: A review of hail science and hail suppression. Meteor. Monogr. (in press).

6. Burtsev, I. I. 1976. Hail suppression, p. 217-22. In Proc. 2d WMO Scientific Conf. on Weather Modification. Boulder, Colo. Aug. 2-6, 1976.

7. Davis, L. G., & P. W. Mielke, Jr. 1974. Statistical analysis of crop damage and hail day rainfall. Annual Report to Lowveld Tobacco Cooperative, Nelspruit, So. Africa. Vol. 2, 1973/74 (Colorado International Corp., Boulder, Colo.) 49 p.

8. Changnon, S. A., Jr. 1977. On the status of hail suppression. Bull. Am. Meteor. Soc. 58 (in press).

9. Atlas, D. 1977. The paradox of hail suppression. Science. 195: 139-45.

10. Schmid, P. 1967. On Grossversuch III, a randomized hail suppression experiment in Switzerland. Proc. Fifth Berkeley Symp. on Math. Statis. & Probability 5:141-60.

11. Grandoso, H. N., & J. V. Iribarne. 1963. Evaluation of the first three years in a hail prevention experiment in Mendoza (Argentina). J. Appl. Math. & Phys. 14:549-53.

12. Long, A. B., E. L. Crow, & A. W. Huggins. 1976. Analysis of the hailfall during 1972-74 in the National Hail Research Experiment, p. 265-72. In Proc. 2d WMO Scientific Conf. on Weather Modification. Boulder, Colo., Aug. 2-6, 1976.

13. Borland, S. W. 1977. Hail suppression: Progress in assessing its benefits and costs. In G. B. Foote & C. A. Knight [eds.] Hail: A review of hail science and hail suppression. Meteor. Mono. (in press).

14. Borland, S. W., & J. J. Snyder. 1975. Effects of weather variables on the prices of great plains croplands. J. Appl. Meteor. 14:686-93.

15. Colorado International Corp. 1974. Final report, weather modification season, 1973/74. Submitted to Lowveld Tobacco Cooperative, Nelspruit, So. Africa (Colorado International Corp., Boulder, Colo.)

16. Taubenfeld, H. J. [ed.]. 1970. Controlling the weather: A study of law and regulatory processes. Dunnellen, New York, 275 p.

17. Fleagle, R. G., J. A. Crutchfield, R. W. Johnson, & M. F. Abdo. 1974. Weather modification in the public interest. Univ. of Washington Press, Seattle. 88 p.

18. Davis., R. J. 1974. Weather modification litigation and statutes, p. 767-86. *In* W. N. Hess [ed.]. Weather and climate modification. John Wiley & Sons, New York.

19. Haas, J. E. 1974. Sociological aspects of weather modification, p. 787-811. *In* W. N. Hess [ed.]. Weather and climate modification. John Wiley & Sons, New York.

20. Danielsen, E. F. 1977. Inherent difficulties in hail probability prediction. *In* G. B. Foote & C. A. Knight [ed.]. Hail: A review of hail science and hail suppression. Meteor. Mono. (in press).

Open Discussion

MILTON KATZ: Can we learn to anticipate side effects of hail suppression, especially changes in the amount of precipitation?

JOHN FIROR: The Great Plains tends to be drier than farmers would prefer, and changes in precipitation affect crops more severely than do changes in hail. A 20% reduction in damaging hail would be completely offset by a 5% reduction in rainfall during the growing season. The amount of water that passes over the Great Plains is many times the amount that falls to the ground. It is not a situation, therefore, of having a certain amount of water that is divided, say, between Colorado and Kansas. The combined amount that Colorado and Kansas receive is only a small fraction of the water that passes overhead in the atmosphere. It might be possible to increase the efficiency of rainfall by modification efforts without depriving anyone else of their precipitation. But the question of distribution of rainfall, particularly in downwind areas, is very complex. The evidence concerning side effects will accumulate much more slowly than will evidence about direct effects, because they are even less certain to occur and we must search for a greater variety of effects.

J. D. NYHART: Apparently we will rely primarily on public funding for research and development in the foreseeable future. Do we know how many persons actually make a living from weather modification? Are there many businesses totally within the private sector?

JOHN FIROR: The United States now exports technology for hail suppression and does so within the private sector. Companies are engaged in hail suppression activities in South Africa and have worked in Kenya and elsewhere. A number of private organizations offer hail suppression services within the United States.

LEWIS GRANT: It is interesting to observe that over the last five to ten years we have experienced generally favorable weather from the standpoint of agriculture, and still there has been a lot of commercial weather modification activity. Now we are entering a drought period, just as we were in the 1950s, which was accompanied by a large demand for weather modifiers. At one time during that dry period, about 40% of the western United States was subject to commercial weather modification activities. The commercial demand for these services now is rising again very rapidly, more for precipitation augmentation than for hail suppression, but this is not necessarily compatible with the requirements of field research. Much of the weather modification research cannot be done in the laboratory, and the field research, to the extent feasible, should be concentrated on questions raised by commercial activities. The demand for commercial weather modification probably will continue to increase in proportion to the severity of the drought.

CHARLES COOPER: There is quite a difference between hail suppression and precipitation augmentation when determining whether the marketplace is a reliable gauge of effectiveness. Some farmers in areas that receive heavy hailstorms are insured and thus are protected against crop damage. It is the insurance companies that have the option of waiting for natural events or of investing in hail suppression activities. It would appear to be in their economic interests to do so, but this would not matter to farmers unless it might result in some alteration of insurance premiums. However, if farmers felt that the indirect effects, such as a change in precipitation amount or pattern, outweighed the possible slight reduction in those premiums, they might exert political pressure to prevent modification. There are many more potential customers for weather modification designed to increase precipitation, and this suggests that precipitation augmentation might be a better test of the marketplace.

JOHN FIROR: That's a relevant observation. Tobacco growers in a region of South Africa contributed to a mutual fund that both would protect their crop by hiring a company from the United States to suppress hail and would pay the growers for hail damage. We should be able to ascertain over a long term whether the amount they paid is more or less than hail insurance premiums would be. This, of course, assumes freedom from political constraints and other factors that preclude valid economic comparisons. We need a long period for these assessments, because the damage in any one year may be zero or may be more than four times the average annual damage. The natural variation is so great that we cannot make valid assessments in any one year.

Let me also relate two conversations I have had with groups with experience hiring weather modifiers. I chided a friend of mine who was on the board of directors of a ski area by pointing out that a "modifier" the organization hired did not know what he was doing. My friend responded with the observation that the weather modifier charged $20,000, which is a pretty good bet for a business with an annual budget of several million dollars and an absolute requirement for snow. He was saying, in effect, that as a responsible member of the board of directors he had to make the investment even if the probability of success was low.

In a remarkably similar conversation with a tea grower in Kenya, who was a member of a joint committee composed of growers that had hired a weather modifier to suppress hail, I inquired whether their figures showed that the hail suppression techniques were effective. These companies keep production records for every field and every field is picked on a 10-day cycle, so their data are fairly extensive. He reported that his impression was that hail damage actually had increased, but that his committee was dedicated to helping the growers and "had to do something." The psychological tendency to invest relatively small amounts of money for disproportionately large returns, even with only a low probability of realizing them, must be considered when analyzing these experiences as market tests.

ARNETT DENNIS: Practical use of weather modification will increase but certainly not at a great rate. Indeed, the industry will grow slowly even in areas where benefit/cost ratios are greatest. As increased experience allows better assessments of benefits and costs, the commercial operators will gradually shift from the most favorable toward the more risky applications in terms of financial returns to them.

Some commercial hail suppression operations have continued in this country for many years. For example, one in southwestern North Dakota has been in operation for 15 years. Commercial hail suppression projects have been discontinued due mostly to perceived undesirable side effects rather than to lack of agreement about effectiveness. The Blue Ridge Project in West Virginia, Maryland, and Pennsylvania is an outstanding example of public perceptions about possible loss of precipitation. Some of the directors of Citizens Against Cloud Seeding, the organization partially responsible for the discontinuance of the South Dakota program, told me they had no argument at all with hail suppression but based their position upon the potential loss of rain. What some persons refer to as hypothetical suppression operations are not really hypothetical, and we will have some rather ambitious ones in this country in the near future.

LOUIS BATTAN: In response to the comments about effectiveness being measured in the marketplace, it is interesting to note what has happened in the U.S.S.R. In the early 1960s, the Soviets began seeding clouds with artillery over vineyards in southern parts of the Soviet Union and claimed that they were reducing damaging hail by from 60% to 90%. In 1964 they were seeding about 200,000 acres (80,000 hectares), and by 1969 they were seeding over 6 million acres (2.4 million hectares), also claiming a reduction in damaging hail of from 60% to 90%, with benefit/cost ratios ranging from 10 to 20. In a 1974 article in *Izvestiya,* Yu Sedunov, then Director of the Institute of Experimental Meteorology, reported that Soviet operators were seeding for hail suppression over 10 million acres (4 million hectares) of agricultural lands, again claiming a reduction of hail damage ranging from 70% to 95%. Nowhere else in the world has anyone claimed such favorable results. An experiment is being organized in Switzerland that is supposed to approximate as nearly as possible the Soviet experiments, so we might have some verification or refutation within several years.

JOHN FIROR: I conclude by reiterating that atmospheric scientists are study-ing very complex phenomena and make few claims for certain success. I say this with some fervor because an article in the latest issue of a national magazine describes the National Center for Atmospheric Research in these words in its opening paragraph:

> High above the city of Boulder, Colorado, is an awesome build-ing, poised like an aerie on a massive upthrusting of Cenozoic rock, in which scientists are plotting the control of our weather.[1]

MODERATOR: That is precisely the type of communication with the public that we hope to prevent through increased recognition of the uncertainties and through greater cooperation among lawyers and scientists to overcome them. This sort of fanciful misrepresentation will be seen frequently unless we increase the public's awareness of our activities.

We will reconvene this evening.

Note

1. Wilkins, K. A. 1976. No one owns the rain. Field & Stream 80(11): 30-31, 122-25.

The Yuba City Episode
in Weather Modification

Dean E. Mann

Department of Political Science
University of California, Santa Barbara
Santa Barbara, California 93106

Introduction

Politics, broadly defined, is the process by which society adjusts to new circumstances, many of which are results of technological change. The political process involves the actions of legislatures, administrative agencies, and courts, and the complex interactions among them.

Our atmosphere is an obvious example of a common property. At least potentially, the action of one person can detrimentally affect others, particularly if the resource is used to its limit or if there is competition for it. Unfortunately, rights to common property or to its use often are ill defined. What are individual and collective rights with respect to the atmosphere? To what quantity and quality is an individual or the collectivity entitled? How pure must the air be? Does an individual or does society have rights to a stated level of visibility, to a given amount of atmospheric moisture, or to certain prevailing winds?

The issues may be posed more appropriately in terms of society's interests. What is society willing to purchase in the way of atmospheric quality or atmospheric resource development? In evaluating the present status of weather modification, its prospects and problems, we may be asking whether society has an obligation to develop the technology—that is, to guarantee specific benefits from the atmosphere—and thereby create a new set of legal rights in atmospheric resources.

Many forces will contribute to the development of weather modification policy. Specific incidents may become precedents for adopting certain policies, or at least furnish justification for doing so. Cloud-seeding activity and resulting litigation about the Yuba City flood probably will help determine future policies with respect to weather modification.

100

The Yuba City Flood

In late December 1955, the northern and central portions of California experienced one of the most devastating storms in the state's recorded history. The storm covered an area of 100,000 square miles (259,000 km^2), about 60% of the area of the state. Nearly 1,000,000 acres (404,800 hectares) of land were inundated, including many highly developed communities such as Stockton, Santa Cruz, Fresno, and Santa Rosa. Estimated direct losses exceeded $200 million, with untold indirect losses. Sixty-four lives were lost and communication and transportation lines were severed over wide areas.

Nowhere were the results of this storm more tragic than in the area of Yuba City, lying at the confluence of the Feather and Yuba Rivers. Confident that the levee system would protect them, the residents did not evacuate and therefore received the full force of the river when the 8-meter levees collapsed and flood waters flowed over 100,000 acres (40,480 hectares), causing damages on the order of $65 million. Thirty-seven persons perished, 3,227 individuals were injured, 467 homes were totally destroyed and 5,745 were damaged, and 8,500 families suffered losses.

Not unnaturally, the community undertook some soul-searching with regard to individual and collective responsibility for the calamity. A grand jury organized in January inquired into responsibility for levee maintenance, flood warnings, and remedial action once the flood had occurred. Nearly three months later, it distributed the blame among three members of the county board of supervisors, five members of the city council, three commissioners of the city levee district, and the state division of water resources. The criticisms were founded on charges of failing to give adequate warnings of the danger, to take corrective action with regard to faulty levees, to interpret warnings correctly, and to provide effective leadership.

In the search for someone to blame, some who suffered damage consulted attorneys, who generally advised them that they had little chance of recovery because the defendants would almost certainly be public agencies. To sue the local levee district would be like suing themselves and to sue the state agency would be difficult because the doctrine of sovereign immunity protects the state.

A committee of the Yuba-Sutter Bar Association investigated the legal grounds for a suit and concluded pessimistically: "It was the consensus . . . [of] the local group that grounds sufficiently strong to assure success in any such litigation in the courts are not apparent." A number of parties who suffered damage were not convinced, however, and through attorneys in the firm of Goldstein, Barceloux & Goldstein they filed suit in 1958 for damages of unspecified amounts in the superior courts of Sutter and Yuba counties against the State of California, Pacific Gas and Electric Company

(PG&E), and North American Weather Consultants (NAWC). Subsequent suits against the federal government because of alleged responsibility of the Army Corps of Engineers in design and construction of the levee system were dismissed.

Cloud-Seeding Activities

PG&E established a meteorology department as early as 1937, the first utility to do so. Its principal interest then was predicting weather to help estimate demand for natural gas. PG&E first employed NAWC to undertake cloud-seeding activities in 1953. This original project quickly expanded over a larger area by 1955 with the goal of increasing snowpack to increase the water supply for PG&E generators downstream.

NAWC and PG&E were reasonably well convinced that their efforts were successful. At the conclusion of the 1955-56 season, PG&E concluded that precipitation in some target watersheds increased an average of 2.9 inches (7.4 cm), or 28.6% above normal, that runoff increased by more than 97,000 acre-feet (120,000,000 m^3), and that the return was $305,842 on an investment of $58,300.

NAWC was reputed to be one of the most responsible and competent firms in the business, with extensive operating experience in the western states for a number of public utilities.

Weather Conditions Before the Flood

Rainfall in December 1955 broke many existing records for northern California, and precipitation during the first half of the month, according to the Weather Bureau, "soaked the soil, filled the streams, and prepared the scene for destructive flooding."

Strong southwesterly winds plus a moderate amount of tropical moisture produced very heavy rains from December 16 to 20 which were followed the next two days by even stronger southwesterly winds and increased amounts of warm moist air. The relatively high temperatures associated with these storms melted snow in the Sierras, with as much as 30 inches (76 cm) of snow melting in three days.

The PG&E meteorology department watched these changes in the weather closely. Their silver iodide generators were operating during the early storm activity but were turned off on December 19 after meteorologists predicted that the storm system after December 20 would be just as severe as before.

They had two reasons for turning off the generators. First, they were convinced that silver iodide particles in warm storms were almost totally ineffective in increasing precipitation because of the difficulty in transporting them to the necessary elevations. The $-5°C$ temperature at which the silver iodide becomes effective was at too high an elevation to be reached from ground generators. Second, they were concerned about the possibility of being accused of contributing to the severity of the storm and perhaps being sued for damages. They accurately foresaw the severity of the oncoming storm and acted to avoid even the appearance of being responsible for it.

Case for the Plaintiffs

The attorneys for the plaintiffs were not aware of the cloud-seeding activities of PG&E until some time in 1956-57, when one of them, Reginald Watt, inadvertently learned the general nature of PG&E and NAWC activities at a luncheon conversation while in another city. Perceiving a possible relationship between this cloud seeding and the Yuba City flood, he sent a student to the area, who spoke with the generator operators, took pictures of their equipment, and learned that PG&E was responsible for the cloud seeding.

Watt immediately started to inform himself about meteorology and weather modification. His search led him to Jack Hubbard, who had academic and practical experience, and who provided Watt with a technical report on whether the activities of PG&E and NAWC could have caused the flooding. While Hubbard appeared to believe that the PG&E generators did increase precipitation, he advised Watt that it would be difficult to isolate this from precipitation caused by natural nuclei, difficult to prove the location of the precipitation, and difficult to overcome a legal situation in which no plaintiff had ever won a case.

Logic dictated that counsel for plaintiffs include among the defendants all persons or interests that might have contributed to the disaster, and to have excluded a potential contributor to the flood, particularly one as capable of making recompense as PG&E, might have resulted in charges of dereliction of duty in protecting the clients. As a result, suit was filed in Sutter County in September 1958, naming the State of California, PG&E, and NAWC as defendants. Eight causes of action concerning the design, construction, and maintenance of the levee system involved only the state as defendant.

The ninth cause of action alleged that PG&E operated artificial rain-making equipment at ten stations in the Sierra Nevada Mountains in areas draining into the watercourse where the levees broke and that the "collapse and breaking of the levees was proximately caused or contributed to by the negligent maintenance and operation of the rain-making equipment, and together with the escaping waters proximately resulted in damage to plaintiffs."

The tenth cause of action simply stated that artificial rainmaking was an ultra-hazardous activity, and the eleventh charged PG&E with poor management of some of its dams.

In preparing his case, Watt encountered some serious difficulties. First, his partners in the firm had little confidence in the weather modification aspect of the litigation and wanted Watt to argue this phase without use of expensive testimony. Watt, on the other hand, felt that their chances were reasonably good and that it would be foolhardy to argue this issue without expert witnesses. He was successful in convincing his partners.

The second problem was obtaining expert witnesses who would testify on behalf of his clients. The leading meteorologists in this field already had been retained by the defendants. Moreover, the desire of commercial cloud seeders to avoid this kind of suit in the future made them unlikely prospects indeed.

In 1962 Watt finally relied on Maurice Garbell as his principal technical expert on the effects of cloud seeding. Garbell was an aeronautical engineer by training and experience, but he had acquired considerable familiarity with meteorology through his professional work. He served as a consultant to well-known clients and was a member of the American Meteorological Society, although he was not certified by it as a consulting meteorologist. He evidently had no direct experience with artificial nucleation, although he was familiar with the principles involved.

Case for the Defendants

Bronson, Bronson & McKinnon of San Francisco represented PG&E, and Price, Postel & Parma of Santa Barbara represented NAWC. Principal attorneys were Edward Morris for the former and Clarke Gaines for the latter, but Morris carried the burden of preparing the case because of the disparity in resources of these two defendants.

Like Watt, Morris was not a meteorologist and had to learn about weather modification. He undertook extensive field investigations, conducted cloud-seeding experiments on his own, and consulted many experts. He prepared himself thoroughly and, in fact, later became president of the Weather Control Research Association, an organization of leaders in weather modification.

PG&E and NAWC contended that plaintiffs had not demonstrated that defendants were responsible for the damage to plaintiffs: There was no proof of causation to link the cloud-seeding activities with the damage in the Yuba City area, no evidence with regard to negligence, and no evidence that the artificially induced rain was either the proximate cause or a contributing factor. Further, the defendants argued that whether artificial nucleation of clouds was an ultra-hazardous activity was a question of law, not a matter for courtroom evidence, and that, indeed, no evidence was presented to demonstrate that it should be considered ultra-hazardous.

Plaintiffs claimed that once they demonstrated that PG&E and NAWC had increased the rainfall and runoff during the devastating storms of December 1955, the burden shifted to the defendants to prove the amount of rainfall and runoff and whether PG&E contributed to the breaks in the levees. The defense argued that pouring one bucket of water into the river should not require proof that it did not cause the levees to break.

Pretrial Proceedings

Pretrial proceedings consumed five years before the case actually went to trial in October 1963. Many reasons contributed to the long delay, not all of which were related to the complexity of the factual issues.

Part of the delay was caused by the difficulty of finding a judge to sit on the case. It was generally felt that a trial on the issue of liability would last from six months to two years. California objected to a local judge sitting on the case, as it had a right to do under California law since property within that county was at issue, so it was necessary to look elsewhere. Attorneys for the litigants requested the judicial council of the state to assist in selecting another judge, but it was not until 1960 that the judge who finally heard the case was found.

A second cause for delay was the changing character of California law. California was a major defendant in this suit and much of the plaintiffs' case was directed toward it. Of particular importance was the changing situation with regard to the doctrine of sovereign immunity as applied in actions of inverse condemnation.

The doctrine of sovereign immunity holds that the state may not be sued unless it has agreed to subject itself to such suits. Watt, among others, had attacked this doctrine in cases which had been appealed to the state supreme court, which in 1961, during the pretrial proceedings, finally overturned the doctrine.[1] The court found the doctrine "an anachronism, without rational basis" that existed "only by force of inertia." This decision essentially eliminated a defense that the state had relied upon during the pretrial proceedings.

Inverse condemnation involves situations where the state damages, takes, or otherwise makes private property less useful through the exercise of its lawful powers and then is sued by the property owner for compensation. Moreover, it may be argued that when the state authorizes another party to undertake an activity that results in a taking or damaging of private property, the state may be liable just as if it took or damaged the property itself.

In two cases decided just before the case went to trial, the state supreme court found the state liable for damages to property resulting from its own activities,[2] and from those of private parties done with approval of the state.[3]

The plaintiffs found support in these cases for their view that the state could be held liable for its part in the authorization, construction, maintenance, and operation of the levee system. More importantly for our purposes, the state had granted a license to NAWC for the purpose of engaging in cloud-seeding activities.

Trial by Judge or Jury?

An important consideration for attorneys representing both sides was whether the case should be tried before only a judge or before a jury. Both sides gave some consideration to having a jury during the pretrial phase, but finally decided in favor of the judge.

The principal consideration involved was the competence of a jury to comprehend the complex technical information to be introduced. The defense carefully assessed the 12,000 registered voters in the county and, eliminating those who worked for the state, PG&E, or NAWC, or who were related through family or employment to the plaintiffs, found only 400 potential jurors, of whom only 120 said they could serve for a year. Both sides felt that the education and experience of these citizens did not qualify them to sit on such a difficult case.

Their willingness to dispense with the jury also reflected the attorneys' confidence in Judge John MacMurray, whom they considered to be a strong, even a "fearless," judge.

Another consideration for the plaintiffs was the attitude of the community. Only 150 of the thousands of persons who suffered damage were litigants. The community generally was hostile—in the view of the attorneys—to the suit, partly due to a feeling of futility. Moreover, citizens who had sustained injury but not sued would undoubtedly be represented on the jury and might assume that plaintiffs should not recover damages if they could not.

A third consideration for the plaintiffs was the estimated daily minimum cost of $120 for a jury. Plaintiffs would have to deposit this money in advance, which would constitute an obvious hardship for a one-year trial.

Finally, the law with regard to trying damage suits was changed in California in September 1963. Under the new law, it became possible to try only the question of liability, leaving for later proceedings the extent of damages, assuming plaintiffs demonstrated liability of the defendants. The issue before the court would be liability alone, not the anguish and hardships of plaintiffs, and this would be determined by technical evidence and an understanding of the law. On this issue, plaintiffs felt the judge would be more competent than a jury.

The Pretrial Order

In October 1963 Judge MacMurray issued a formal pretrial order in which he stated the causes of action and the contentions of the plaintiffs and the defendants.

He summarized the following contentions of the defendants: (a) Rainstorms caused the damages, (b) the storm was an act of God, (c) plaintiffs failed to state facts sufficient to constitute a cause of action, (d) an independent contractor of PG&E seeded the clouds (presumably making NAWC culpable but not PG&E), (e) plaintiffs assumed the risk or contributed to the negligent conduct, (f) fire-damaged conditions of the watersheds resulted in partial damage to the levees, and (g) no artificially induced rain fell outside the target watershed during December 15-19.

The order outlined the following issues of concern to PG&E and NAWC: (a) Negligence on the part of PG&E and NAWC, (b) extraordinary force of nature or act of God, (c) strict liability for the conduct of ultra-hazardous operations, (d) contributory negligence of plaintiffs, (e) assumption of risks by plaintiffs, and (f) whether NAWC was an independent contractor in such a relationship as to exonerate PG&E from liability.

The Trial

The assumption of widespread community interest in the trial was not confirmed. During the first three days of the trial in October 1963, only one plaintiff attended each day, and each day it was a different one. There were no spectators.

The local newspaper carried stories on the front page for the first week of the trial, but by late January 1964, when the trial ended, notices generally were relegated to inside pages.

Witness for the Plaintiff

The first and only witness to appear personally for the plaintiffs on the issue of liability from cloud seeding was Maurice Garbell. He occupied the witness stand for a total of eight days, during which he testified on the technical aspects of cloud seeding and the character of the 1955 storms and replied to intensive cross-examinations by Edward Morris.

The testimony focused on how the silver iodide was emitted from the generators, then transported with lower air movements until caught up in convection cells that rapidly lifted it to regions of cooler air where nucleation could begin. The crucial questions, then, were the extent to which such cells were present, the rate of ascent of the silver iodide plume, the elevation at which the required temperatures were found, and the time required for silver iodide to cause moisture to crystallize. The rates at which these processes occurred determined whether the resulting precipitation would have fallen on the target watershed or would have been carried beyond it.

It was necessary for the plaintiffs to show that these processes occurred at a rapid rate with minimal dispersion of the plume as it was transmitted aloft. The witness contended that the crystals ascended in convection cells at the rate of 5 meters a second and reached air temperatures of $-5°C$ at 2,700 meters on December 18 and at 3,000 meters on December 19. At the 2,700-meter level, the crystals would have traveled a maximum of 37 kilometers from one of the generators. He also testified that the nucleation process would begin the moment the crystals attained the necessary temperature.

Garbell functioned as an expert witness on general meteorology as well, stating that the signs of a severe storm were sufficiently clear for PG&E and NAWC to have predicted what the effect of seeding could be and that California should have recognized the possibilities of such storms and the consequent floods.

The defense strategy was to force the plaintiffs to prove the causal relationship from seeding to precipitation to flooding, while casting suspicion on any such relationship and minimizing the effect of NAWC's cloud seeding, even if the silver iodide did reach the elevation necessary for nucleation.

The defense challenged the atmospheric models relied upon by Garbell, and the testimony became very complicated as he and Morris argued the probable rates of ascent, the likely lateral movement of the plume, the atmospheric temperature gradients within the storms, and related scientific issues.

Motion for Nonsuit

At the conclusion of plaintiffs' testimony, PG&E and NAWC moved for a nonsuit on the grounds that the plaintiffs failed to demonstrate legal culpability and that it was unnecessary for defendants to present a positive defense. NAWC had limited financial resources and already had gone through one of the longest pretrials in California's history, recently lost its insurance as a result of this litigation, and also lost considerable business since the case began.

The chances of Judge MacMurray granting the motion were slight, because of the merits of the plaintiffs' case and because of the probability of an appeal, regardless of the outcome. Judge MacMurray appeared eager to ensure that the trial court have a complete record for the appeals court and that the case not be remanded because of judicial error. The motion for nonsuit was denied.

Witnesses for the Defendants

The first defense witness, Vincent Schaefer, testified both on the science of artificial nucleation in general and on the specific scientific questions involved in this litigation. He informed the court that the process of artificial nucleation required from 14 to 21 minutes for the ice crystals to achieve sufficient mass to fall and that in this instance they would have been carried beyond the drainage area that contributed to the flood. He further concluded that that activity of NAWC could not be considered either hazardous or ultrahazardous.

The defendants called their second expert witness, Arnold Court, specifically to cast doubts on the validity of the cloud seeders' claims because of the difficulty of establishing sufficiently controlled conditions. He contended that a longer period of intensive experimentation would be necessary to prove their claims. After reviewing the data upon which PG&E and NAWC relied in evaluating their cloud-seeding activities, he thought it possible to demonstrate that more rain fell without seeding than with seeding and observed that PG&E personnel seemed determined to prove that cloud seeding could produce additional precipitation.

To buttress his case, Morris himself presented calculations to show that the seeding effort could have produced only 572 acre-feet (707,000 cubic meters) at the levee and cited calculations by the Corps of Engineers that showed that a reduction of even 52,000 acre-feet (64.3 million cubic meters) at the time of the break would not have made any difference. Dramatically displaying a small jar of silver iodide crystals before the court, he announced that it equaled the quantity of crystals emitted by all the generators for a period of 50 hours before they were turned off and asserted that it hardly could be considered an element in causing the levee failures.

Morris was prepared to argue, if necessary, that the artificial nucleation actually reduced rather than added to the stream flow at the time of the flood, if it produced any precipitation at all. He would contend that the precipitation would be snow that would increase the snowpack and absorb subsequent precipitation in the form of rain, thus reducing the amount of water in the river at the time of the break. Watt was prepared to counter by contending that earlier rains had sluiced off the snow so that later precipitation fell on bare but saturated ground and rapidly entered the watercourses.

In summing up, Morris pointed to the public policy implications of placing the burden of responsibility on the cloud seeders: "If it is branded by the courts as ultra-hazardous, it will be shunned by research organizations and universities. The insurance industry will undoubtedly refuse to underwrite it and private enterprise will no longer be able to engage in it."

Negotiation of a Settlement

Sometime during the trial, attorneys for each side began a series of discussions that led to final disposition of the case. Just when these negotiations began remains unclear, but the parties generally agree about the motives behind them.

From the standpoint of PG&E, it was important to achieve a favorable decision to avoid an appeal. Its attorneys discussed affairs with the insurance companies that would be called upon to pay any damages against PG&E. A new trial would require more pretrial proceedings, with the probability that plaintiffs would include another project in its list of allegations, thus involving additional expense. Under these circumstances PG&E management and the insurance companies advised negotiation of a settlement. The attorneys for PG&E were convinced that Watt would be willing to settle because they felt that the witnesses for the defendants devastated the plaintiffs' case and that Watt might want to pull back from a weak case in favor of concentrating on the stronger case against the state, which seemed to be going well.

However, Watt believed there was a good chance of winning and that the court probably would rule against both defendants. But he concluded it was not particularly important that it rule against PG&E and NAWC, since his clients could collect damages from the state. He also may have preferred to concentrate his efforts on what appeared to be the better case.

The situation for Watt was complicated by the fact that he could not locate all his clients to obtain their agreement to settle. Therefore, it was necessary to strike an arrangement that would not expose him to charges of not consulting with clients before such a major decision. Both sides negotiated willingly and reached an agreement long before the conclusion of the trial. If PG&E and NAWC won, the plaintiff would not take an appeal within the 30 days permitted by law. If plaintiffs won, however, PG&E and NAWC would be free to take whatever steps they thought necessary to defend their interests. In turn, the defendants agreed to pay the attorney fees for the plaintiffs. This is known in the legal profession as "buying an appeal."

Judge MacMurray, who was aware of the negotiations and agreement, issued the following decision in April 1964:

> Plaintiffs may not recover against the Pacific Gas and Electric Company or North American Weather Consultants as they have failed in their burden of proof. . . .
> [Neither PG&E nor NAWC] produced any significant increase in rainfall or snowfall outside of the Lake Almanor watershed. The effects of cloud seeding were limited to the pre-determined target area which drains only into Lake Almanor. Lake Almanor never spilled at any time before or during the flood; accordingly, any increase produced by cloud seeding was successfully impounded by that Pacific Gas and Electric Company lake.
> The breaking of the levees was neither proximately caused nor contributed to either by the maintenance or by the operation of the artificial rainmaking equipment of any defendant in this lawsuit.[4]

Judge MacMurray ruled against the state concerning its responsibility for the design, construction, and maintenance of the levee system and ruled that damages could be recovered on the legal ground of inverse condemnation: "Plaintiffs may also recover in strict liability against the State of California as strict liability must be imposed when a failure occurs in a levee system constructed and maintained to protect the lives and property of people in the area when such failure occurs under reasonably foreseeable circumstances."

California considered appealing the decision and engaged Edward Morris to investigate how the state had fared in appeals involving liability in water cases.

He found that the state had never won an appeal, so the state decided against one. Plaintiffs originally claimed damages in excess of $13 million; the parties settled for a total of $6.3 million.

Implications for Public Policy

The *Yuba City* case does not allow easy generalizations. Few cases—whether in law, medicine, or political science—delineate a field in its entirety. It may provide insights and new ways of addressing issues; it may rule out certain approaches that had been thought workable; it may lead to new hypotheses about certain phenomena; and it may indicate some order or predictability about certain relationships.

This case is not entirely satisfactory because the full legal implications of artificial nucleation were not explored. The court based its decision on insufficient proof of causation. The mere fact that cloud seeding occurred before a storm caused damage does not prove that the seeding caused the harm. We can only speculate about whether cloud seeding is an ultra-hazardous activity. By implication, at least, the court appeared to say that artificial nucleation of clouds was not an ultra-hazardous activity against which arguments of prudence and reasonable precaution by the seeder would be unavailing.

A further complicating factor was that the alleged weather modifiers were co-defendants with others whose culpability was—at least in retrospect—more readily demonstrable. It is interesting to speculate about the result if PG&E and NAWC had been the only defendants. The 150 plaintiffs, who could never hope for full compensation, were asking for some redress. Would the social consequences of a decision that denied them redress have led to a different conclusion? Courts consider the social consequences of their decisions and it was convenient for this court to have the state included as co-defendent.

Still another difficulty in generalizing from this case is the lack of an appeal on the legal issues. The legal implications of altering the amount of precipitation over an area were not examined thoroughly, as they might have been on appeal.

These reservations about drawing generalizations from this case do not preclude observations about the nature of issues likely to develop concerning weather modification. First, in view of the complexities of water management today, it is uncertain how the law of weather modification will merge with the law of other water resources. Water is managed intensively throughout the country by a diversity of private and public entities—*e.g.,* public utilities, government agencies, private individuals, and local public districts—with different purposes and methods that sometimes conflict. Changing the weather, more specifically the precipitation patterns, to any appreciable extent will affect not only the modifiers but also those who manage dams, channel the water in streams, divert the water for various purposes, alter the ground cover to influence run-off, or in other ways manage water. Future litigation promises to be complex and to involve numerous parties because cloud seeding seldom will be perceived as an isolated activity.

A second observation stems from the ironic situation in which the defendants found themselves in this case. PG&E invested in cloud seeding with hopes of increasing snowpack and appeared to be convinced by results of the NAWC operations that such increases occurred. NAWC existed because its scientists believed that artificial nucleation of clouds could produce appreciable increases in precipitation. Yet they argued in this case that no such results had been obtained. This probably will remain the first line of defense because of the infancy of weather modification. Sufficiently reliable evidence about the effectiveness of weather modification techniques simply does not exist. Satisfying scientists on a statistical basis is easier than demonstrating effectiveness in a specific storm under given conditions.

A third observation relates to the competence of courts to deal with issues of this kind without the benefit of special investigations. Given the complex factual and legal issues that are likely to come before the courts, there seems little doubt that administrative proceedings based on the steady accumulation of technical information and legal expertise would greatly assist in assigning responsibility. Related to this is the question of statutory basis for administrative decision making. The agencies must know the degree of discretion they have with regard to the issuance of licenses and permits. It is obvious that mere publication of information about planned weather modification activities in a newspaper provides virtually no protection for those whose interests may be affected. Even the few persons who would be aware of the projects might not perceive their importance and would be unlikely to act to prevent the operations unless a direct threat were evident. This means in essence that ordinary citizens can protect their interests only by expensive lawsuits after the fact. Legislatures should define clearly the government's responsibility to protect the interests of all citizens.

The public has an equally vital interest in continued experimentation in weather modification. Indeed, there is a strong predisposition in the West in favor of anyone who can augment the water supply, but the weather modifiers share an obvious concern about protecting themselves from litigation. During the course of the *Yuba City* litigation, NAWC lost its insurance, which was subsequently restored. It appears that insurance companies also raised questions about the desirability of PG&E continuing such risky operations.

Insuring those engaged in weather modification may involve no more risks than insuring many other technological enterprises, but the elements of novelty and uncertainty inhibit insurers not familiar with it from providing coverage. Business volume is limited, and there is little history with regard to premiums, losses, and standardization of policies.

The Weather Modification Association (WMA), whose members are those engaged in weather modification activities, has been trying to get an insurance "package" for members who meet certain standards. This would facilitate the acquisition of insurance by the operators and would provide the insurers with confidence about the practice of operators. The threat of expulsion or censure by WMA might promote high professional standards.

The *Yuba City* litigation prompted the acquisition of more reliable evidence on the results of weather modification projects. For example, PG&E meteorologists stated that the litigation caused PG&E to use randomized experiments in accord with prior recommendations of statisticians, and the company has intensified its monitoring program to gather more reliable data about the quantities of water produced by seeding. These are done both to assess the economies of seeding and to provide PG&E with better information should it be sued again.

It is not difficult to imagine a set of circumstances under which a weather modifier might be held liable for damages resulting from cloud seeding. Would this spell the end of weather modification? This is hardly likely. Rather, an adverse decision probably would stimulate even greater attention to standards of performance and monitoring. This in turn would stimulate scientific endeavor rather than retard it, and promote greater cooperation between scientists and lawyers. Otherwise, the potential benefits of weather modification will not be realized.

Editor's Note. For an extended account of the scientific and legal issues surrounding this lawsuit, see the longer article by D. E. Mann, The Yuba City flood: a case study of weather modification litigation, Bull. Am. Meterol. Soc'y 49:690-714 (1968).

Notes

1. Muskopf v. Corning Hosp. Dist., 55 Cal. 2d 211, 359 P.2d 457, 11 Cal. Rptr. 89 (1961).

2. Youngblood v. Los Angeles Flood Control Dist., 56 Cal. 2d 603, 364 P.2d 840, 15 Cal. Rptr. 904 (1961).

3. Frustruck v. Fairfax, 212 Cal. App. 2d 345, 28 Cal. Rptr. 357 (1963).

4. Adams et al. v. California et al., No. 10112 (Sutter County Super. Ct., filed Apr. 6, 1964).

Reports by Working Group Leaders

Social Implications of
Weather Modification

Barbara C. Farhar

Human Ecology Research Services
855 Broadway
Boulder, Colorado 80302

Participants generally agreed that the following issues dominate the social implications of weather modification.

1. Public Pressure for Operational Application

We could not avoid discussing the current drought and its effect on attitudes about weather modification. It is both an incentive and a deterrent to operational programs. Crop failures and water shortages could result in public pressure for precipitation enhancement when the technology still is unready to deal successfully with such major weather phenomena. Unsuccessful intervention or seeding programs that produce few beneficial results, coupled with unrealistic expectations by the public about what seeding could accomplish, might result in public rejection of weather modification. This occurred in South Dakota, culminating in the demise of the South Dakota Weather Modification Program in 1976. Such rejection can impede further application of a potentially valuable technology.

The timing of this intervention may be important. If a drought cycle is predicted, it might be wise policy to avoid introduction of the technology while the drought is becoming more severe. When more favorable weather cycles are predicted—*i.e.,* with adequate precipitation—weather modification projects more likely would be perceived favorably. Even if they fail, consequences to those in the target area would be less severe.

An issue of continuing concern is how to protect society by maintaining professional standards for weather modification operations. If public pressure for operational precipitation enhancement during droughts continues, we should monitor and evaluate these efforts. Operators should be required to collect data that would be made available for evaluation by independent organizations.

2. Scientifically Uncertain State of the Art

The state of current knowledge about summer precipitation augmentation and hail suppression can be summarized as uncertain. Unfortunately, no sufficient long-range research programs exist to overcome the uncertainty. This lack, together with the likelihood of public pressure for operational programs, promotes a policy of reacting to crises rather than preparing for them. Some participants in our group felt strongly that operational programs should not occur until better scientific knowledge exists, although they agreed that such a prohibition was politically unlikely.

In addition to needing better understanding of basic atmospheric processes and of modification effects on them, we need to address the environmental, social, and legal uncertainties of weather modification.

3. Inadequate Public Policy

As noted above, the group felt that current policy on weather modification could be characterized as reactive rather than proactive. No policy analyses have been conducted to show how weather modification coincides with or contributes to broad national goals, such as improving human health, maintaining national security, providing sufficient energy supplies, enhancing environmental quality, and producing food and fiber. The group agreed that weather modification should be considered as a means to one or more of these ends. It is not wise national policy to react to an individual crisis like the drought. Rather, we should promote a coordinated and well-planned capability to act.

4. Need for Adequate Decision-making Mechanisms and Regulation

Members of the group readily agreed that adequate decision-making mechanisms and institutional arrangements for using weather modification to promote the public good are lacking. As one member stated, "All too often, controversial weather modification activities are subject to public debate that consists of loud charges and counter-charges. This lack of sound decision-making procedure almost invariably leads to legislative prohibitions or impediments and to pressures from business." For example, in one documented case, a major purchaser of barley threatened to reduce purchases by 10% annually unless a weather modification program was allowed to continue in the growing area. [1,2] These certainly are not scientific decisions based upon technology assessments and the overall public good.

Possible public pressure to initiate operational programs with a scientifically unprepared technology raises questions about regulating weather modifiers or projects to protect the public. Other issues that might require regulation are allocation of atmospheric water, liability for damages, and compensation of persons harmed. Management of an uncertain technology is at least as important as management of a certain technology.

Regulation of weather modification occurs primarily at the state level, and the current diversity in existing state laws might be beneficial because so little is known about how this activity is best regulated. We should learn from the diverse approaches tried by various states.

We discussed at length the division of regulatory responsibilities among local, state, and federal government. Governmental intervention is uncommon in scientific affairs but is common in technological ones. We must be careful in defining what we want from local, state, and federal government and should not overlook international agreements. We recognize that federal regulation tends to accompany federal funding.

5. Social Considerations

Information dissemination and educational programs for the public and the appropriate decision makers are inadequate. Publicly supported programs should include unbiased information that explains the project, thus lessening the chance of disillusionment with weather modification. Disappointment will inevitably follow unrealistically high expectations and will negatively affect research programs as well as operational attempts. Many legislators and administrators with regulatory responsibilities currently lack adequate information about weather modification.

Credibility of advocates for weather modification is related to the scientific uncertainties of the technology, to the lack of public education, and to the unwillingness of many scientists to speak candidly about the statistical nature of results from weather modification research. The general belief that members of the public are more willing to accept weather modification if they know more about it is not supported by research findings.[3] While some opposition may be based upon inaccurate ideas or assumptions, the basic issue concerns the distribution of risks and benefits. Even if a technology is almost certain to produce given effects, only some members of the public will want them while others will vehemently resist.

Conflicts about what weather conditions are desirable—heterogeneity of weather needs—is yet another problem. What one segment of the public wants or needs is not necessarily what others in the region want or need. For example, Colorado barley growers might want a dry period for harvesting at the same time that ranchers want precipitation for the range. This underscores the requirement for an institutional arrangement whereby "winners" from a modification project somehow compensate "losers."

In general, the future issues we identified are extensions of the current ones. Cycles of public pressure for cloud seeding in response to crisis situations, followed by relatively little public interest in the technology, are likely unless policies are formulated to preclude them. Uncertainty about effects and unrealistic expectations will continue until we have adequate scientific knowledge and effective public education programs. All current issues will become more critical in the future, and failure to adopt a comprehensive national policy—as distinguished from regulation alone—will exacerbate them.

The group recommends the following actions to promote intelligent development of a potentially beneficial technology.

1. An analysis of national policy regarding weather modification and how it relates to national goals should be conducted, and the priorities for guiding the technologies should be firmly established. This policy should include all federal interests, such as research and development, operational, and regulatory. This recommendation does not imply that the federal government should exercise total control over weather modification but does encourage a federal commitment to place weather modification in proper context relative to national goals.

2. Regional seminars, perhaps sponsored by the AAAS and the ABA, should be convened to bring together experts on weather modification and private and public decision makers to exchange information about the current needs and trends.

3. Evaluation of operational projects should be mandated and coordinated by the federal government. One agency should be delegated the clear authority needed to preclude inadequate performance caused by interagency rivalries. The specified entity—perhaps a National Climate Program—would serve as data repository, would perform independent analyses of project effects, and would sponsor educational activities to foster cooperative efforts. The "ships of opportunity" concept might be used to establish the necessary data base. These are naval and commercial vessels equipped with oceanographic test devices for data collection in areas not normally visited by regular research vessels. Similarly, commercial and governmental organizations could cooperate to increase the information needed to plan weather modification activities and to monitor the results.

4. Basic and applied research in atmospheric sciences that advances our knowledge of the effects of weather modification and research in the associated environmental,[4] societal, and legal areas should be accelerated by private and public organizations.

Notes

1. Lansford, H. 1973. Weather modification: the public will decide. Bull. Amer. Meteorol. Soc'y 54:658-60.

2. Farhar, B. C. 1975. Weather modification in the United States: A socio-political analysis. Ph.D. Thesis. Univ. Colorado. 400 p. (Diss. Abstr. 75-23,598).

3. Farhar, B. C., & J. Mewes. 1976. Social acceptance of weather modification: The emergent South Dakota controversy. Monogr. No. 23. Institute Behaviorial Science, Univ. Colorado.

4. Cooper, C. F., & W. C. Jolly. 1970. Ecological effects of silver iodide and other weather modification agents: a review. Water Resources Res. 6:88-98.

Institutional Relations and
Dissemination of Information
Dean E. Mann

Department of Political Science
University of California, Santa Barbara
Santa Barbara, California 93106

We explored the basic premises for approaching institutional questions and concerned ourselves with how institutions of various types could be structured to achieve national goals. We heard some concern about the unfortunate consequences that might follow a series of relatively minor incorrect decisions by courts or legislatures without consideration of the broader implications of how this new technology might be used most profitably. We also discussed how weather modification activities could promote national policies and, even more basically, how these national policies are defined. We felt that the smaller decisions should be made with full appreciation of the larger rationales.

As a general statement, we sought to ascertain how weather modification might serve broad general purposes. We all recognize that it is not an end in itself and that it must be made to serve societal interests. We must examine these interests as a fundamental feature of any decision about the appropriate institutional arrangements for either promoting or constraining weather modification operations. Because the purpose of this conference is to deal with how scientists and lawyers should work together—with weather modification serving as an illustration of the challenges—we considered existing instances of cooperation and let them lead to more specific considerations of how to improve cooperation with respect to weather modification. For one example, scientists or technologists seek the advice of lawyers about the consequences of their proposed activities and about handling their personal or commercial affairs. Scientists and technicians also go to lawyers to obtain damages, to prepare contracts, and otherwise to participate in an attorney-client relationship. In the academic setting, lawyers and scientists help each other by conducting research in areas of mutual interest.

There was a clear expression of belief in our working group about the need to bring lawyers and scientists together in even more informal settings than we have here today. There are a number of ways to do this. One is the regional conference forum that Barbara Farhar suggested on behalf of her working group, in which lawyers and scientists might be asked to focus on weather modification where the informal surroundings would encourage discussions on other areas as well. We also think that opportunities exist at universities to cross-fertilize the minds of law students and students in the scientific departments. We strongly encourage this type of educational program, both as a classroom undertaking and as a possibility for collaborative research or action projects. This would enable the students to recognize both the opportunities and the constraints under which other professions work. Many employers —including the federal government—provide programs for middle-management employees to broaden their awareness of appropriate topics central or related to their employment. These programs are appropriate models for designing ones to bring scientists and lawyers together. This could be an extraordinarily beneficial program throughout governmental agencies.

The AAAS and other professional organizations have congressional fellowship programs by which scientists are brought into daily contact with congressional staff members for first-hand experience in the legislative process. We also discussed opportunities for funding legal research directly related to scientific problems. The National Science Foundation finances various legal studies concerning use of solar energy, modification of weather, and related subjects. We emphasize the need for legal research that parallels ongoing activities in the sciences, particularly with respect to weather modification. The time lag in society between paying attention to new scientific or technological achievements and to resulting legal or institutional matters is estimated as being from 7 to 20 years. We strongly suggest that this gap needs to be reduced to provide better institutional controls for dealing with the consequences of scientific and technological progress. We need first to examine the barriers that our legal system presents to effective weather modification programs and then to examine how these can be modified to encourage responsible research activities.

One of the most serious impediments to continued research in weather modification is the prospect that anyone who engages in this activity will be sued for damages. This raises immediately the question of indemnification, and we concluded that the probability of Congress's passing indemnification laws is relatively low. Nevertheless, we think it desirable for lawyers to consider the questions surrounding various systems of providing compensation that will allow weather modification research to continue.

These issues, of course, should not be examined individually but in proper context with other kinds of legislation and other institutional programs.

It should come as no surprise at this point in the conference that we all agree on the necessity of field research to advance meteorological knowledge. We should not expect rapid progress from laboratory and computer studies alone. These large-scale research efforts and operational programs should be accompanied by investigations designed to resolve some of the social and legal problems as well as the physical ones. The present fragmentation is slowly being overcome, but we need to make a greater effort to do so.

These considerations led to discussions about institutional frameworks to provide effective mechanisms for coordinated efforts. The most general conclusion is that a special commission for assessing what we know and what we should know before proceeding with large-scale weather modification projects might be created, for either short-term or long-term purposes. Similar commissions or organizations that might be used as models were designed with five-year lifetimes and, although we are not certain this is the proper period, it does seem about right. Perhaps it would be possible to create one for a specified number of years. This organization would not deprive the mission-oriented agencies of their responsibilities and, indeed, it would be necessary and appropriate for these agencies to assume full responsibility for programmatic tasks. We do not wish to burden the agencies with another layer of bureaucracy but suggest that a commission of this sort might serve a needed role in coordinating efforts within the agencies. Interstate activities require a federal presence of some sort.

Due to the multiplicity of interests involved in weather modification activities, we should avoid the tendency to concentrate too much responsibility within a single agency. This would avoid the conflicts of interest that characterized the Atomic Energy Commission until its division into the Nuclear Regulatory Commission and the Energy Research and Development Administration.

Before a federal coordinating role can be effective, however, we need a strong statement of federal policy. However, as Barbara Farhar also pointed out, we presently have none. The special commission might draft an explicit statement of policy that would guide courts and legislatures as they endeavor to balance the various interests of society.

A clear statement of national goals or objectives—or a statement of how weather modification should be used to achieve other national goals or objectives—would help overcome the uncertainties surrounding allocation of responsibilities among various levels of government and private entrepreneurs. We came to no clear conclusion about how to approach this but suggest that over time, as we gain research and operational experience, these policy objectives will emerge. However, a special effort must be made to articulate them if they do not emerge in the natural course of events.

A coordinated approach that involves all relevant interests would also help in the commercialization phase. We might in the very near future be called on to devise appropriate arrangements by which the federal government facilitates the adoption of weather modification technology with appropriate constraints. We need to consider prospectively the role of private enterprise in this developing technology. We must think about designing appropriate institutions for research and appropriate institutions for regulation.

Finally, we need to pay more attention to inadvertent weather modification because it will be increasingly important in shaping our social values, and thus raise proportionately greater legal and scientific questions. It might be wise to distinguish carefully the advertent from the inadvertent effects for the public.

DAVID ROSE: Did you consider the institutional arrangements for managing watersheds and airsheds, particularly where these two occupy the same geographical area?

DEAN MANN: This is closely related to the general question of how weather modification contributes to national goals. Clearly, one of these is ample production of food and fiber, and this quite obviously requires effective management of our water resources. No, we did not discuss the watershed or airshed management problems specifically.

JAMES SMITH: We keep talking about food production, but I think it is important to realize that much of the weather modification activity over the past 25 years has been undertaken to provide water for hydroelectric power production in the West, and to a lesser degree for municipal and industrial water supplies.

MODERATOR: That's an important point, and I think that none of the working group leaders overlooked the diversity of national objectives.

Risk/Benefit Analyses

Stewart W. Borland

Economics Branch
Agriculture Canada
Sir John Carling Building
Ottawa, Ontario K1A 0C5

We treated the term "risk/benefit analysis" (RBA) as if it included the entire set of quantitatively oriented methods also referred to as "cost/benefit analysis," "decision analysis," or "impact analysis." Such a broad definition was adopted because these methods share many features and seek to answer the same general question: "What are the implications for society of certain policy decisions concerning the use of a developing technology?"

We agreed on the following agenda for our group:

1. Exchange information about the nature and application of RBA in the field of weather modification, and
2. Attempt to answer the question, "How can the methodology of RBA help the law to channel the development of this technology in a constructive manner?"

The concept of a risk/reward ratio and the observed differences between people's willingness to accept risks that are voluntarily assumed versus those that are perceived as being imposed by society were reviewed by reference to Starr's article, "Social Benefit Versus Technological Risk."[1] We then discussed two specific examples of applying the RBA to weather modification. The first was the study by the Stanford Research Institute on the decision to seed hurricanes,[2] and the second was the controversy about the possible role of cloud seeding in 1972 in the Rapid City flood.[3,4] In both instances, the RBA was performed within the framework of decision analysis, so that the fundamental uncertainties about the physical processes were made explicit.

The importance of the interaction between the geophysical event and the relevant population at risk was highlighted by Don Friedman's description of computer simulation of natural hazard occurrences. This combines the physical severity of the event with the spatial distribution of the people and property exposed to it as a way to predict realistic estimates of loss. Details of this approach are explained in an appendix to this report.

Workshop participants felt that the following problems have arisen in previous attempts to apply RBA to the social issues generated by developing technologies:

1. Ignoring or not understanding the influences of "soft" variables, such as shifts in distribution of income, aesthetic effects, and interferences with moral or religious beliefs. Because the techniques of RBA are not yet capable of dealing in an entirely adequate manner with such incommensurables, the resulting conclusions suffered a serious loss of credibility with the public.
2. Emphasizing the importance of outcomes and effects, thereby neglecting the intrinsic value of the procedures used. The means to a stated end have their own values that must be considered.
3. Overlooking the need to insist upon application of RBA to technological issues in sufficient depth and at appropriate times.
4. Proceeding without adequate mechanisms to deal with situations in which new technological developments alter accepted probabilities and undermine the basis for current precedents.
5. Financing an emerging technology without provision for the costs imposed by the level of monitoring required to provide satisfactory data with which to perform sound RBA.
6. Assessing inadequately the utility of alternative outcomes. These depend almost entirely on whose preferences are given the most weight and on what assumptions are made regarding the changing nature of those preferences over time.
7. Establishing the point at which the level of risk is no more than "mere speculation," a term used in rendering the 1950 verdict in *Slutsky v. City of New York*.[5]
8. Agreeing on whether "overwhelming confidence" in the harmlessness of modification is needed if public safety or environmental irreversibilities are involved.

Workshop participants reached general agreement on the following points:

1. Notwithstanding their defects and limitations, no satisfactory alternatives exist to the use of policy-analytic methods for advising decision makers about the development and application of weather modification.

2. Thoughtful performance of RBA and similar analyses does lead to better understanding of the limitations of available data and of the priorities for future research.

3. Policy-analytic methods are extremely useful in hastening the convergence of initially differing judgments as information accumulates over time.

4. Confidence in the quantitative estimates is fundamental to their usefulness to decision makers.

5. It is important to distinguish between the following situations in deciding which levels of risk are considered acceptable: (a) Voluntarily assumed exposure versus involuntary exposure, (b) risks associated with research versus those generated by commercial or operational programs, and (c) risks associated with irreversible environmental or human health effects versus those involved with temporary ecological imbalances or other effects.

6. Some types of weather modification should be reserved for crisis situations—*e.g.*, hurricane diversion—and then used only if public demand is unequivocal.

7. When net social benefits are clear, society often does not compensate individuals who may suffer private losses due to implementation of a program and should not be bound always to do so in the case of weather modification.

We concluded our session by drafting these recommendations for action:

1. Policy-analytic methods should be initiated in the design stages of research and operational weather modification programs to ensure that a broad range of issues and interests will be considered in time to influence the planning of activities.

2. In applying RBA to weather modification, consideration should be given to a wider range of alternative means of attaining the desired objectives.

3. Attention should be directed to bridging the gap between numerical expressions of probability—as in science—and verbal expressions of probability—as in jurisprudence. Expressing the relative accuracy of estimates derived by RBA—*e.g.*, through display of statistical confidence limits surrounding the estimates—might ease the difficulty of comparing them.

4. In view of the present extreme uncertainty about potential physical effects, no significant weather modification program should proceed without some type of public referendum.

5. The present system of reporting weather modification activities should be examined carefully to ensure that it provides the data required for the desired quantitative analyses. Some minimum level of project monitoring should be an integral part of the reporting system.

APPENDIX

Natural Hazard Assessment Using Computer Simulation
Don G. Friedman

Travelers Insurance Company
Hartford, Connecticut 06115

Natural hazard simulation provides a quantitative means of estimating the effect of geophysical events upon certain segments of the United States economy. This approach also can be used to assess possible implications of large-scale weather modification on the damaging effects of severe storms, such as hurricanes and thunderstorms.

Every section of the United States is affected by one or more natural hazards, including coastal and inland flooding, hurricanes, tornadoes, hailstorms, and windstorms. Each year hundreds of lives are lost and billions of dollars in damages occur. The number and severity of natural disasters have increased sharply in recent years even though we have no conclusive evidence that present meteorological events are more severe than in the past. Population growth coupled with greater susceptibility to natural hazards increases the potential losses.

Insurance is one means of reducing the detrimental effects. The possible impact of future disasters upon an insurer is difficult to assess using only past loss experience because of peculiarities in loss-producing characteristics of natural events. An approach involving computer simulation techniques provides one way to make this assessment.

Using this approach, a measure of present and future losses due to natural hazards in the United States can be developed. In the analysis, four factors interact to determine the magnitude of potential losses due to an event of given intensity and location. These are (a) a natural hazard generator, (b) the modifying effect of local conditions, (c) the geographical array of populations and/or structures at risk, and (d) the vulnerability of these populations-at-risk to loss when an event of given severity occurs.

The natural hazard generator determines the frequency and severity of geophysical events in different sections of the United States and generates a geographical pattern of severity associated with those events. The size and shape of the pattern depends on the input characteristics of the simulated natural hazard. The measure of severity can be expressed in terms of maximum wind speed associated with the passage of a hurricane or of other damaging features of a storm. Modification of the pattern due to local conditions—such as urban versus rural exposure to wind—is included in the analysis.

Population-at-risk specifies the number and geographic distribution of persons or structures in the United States. An 85,000-point grid system represents the land area of the contiguous 48 states, with each grid area being approximately the size of a "township." The geographical distribution of the 220 million persons and the 50 million single-family dwellings in the United States are examples of populations currently used in the simulations.

The overlapping and resulting interaction of a geophysical event's severity pattern with the geographical array of the populations-at-risk determines the magnitude of the potential loss. This potential can vary from very low, as when a weak storm affects a sparsely populated area of the United States, to very high, as when a hurricane passes through or near an urban area. A slight change in position of the severity pattern relative to the population-at-risk array can drastically change the potential loss. When the potential is large, it indicates the likelihood of experiencing a natural disaster. Many different overlappings are possible. One cannot afford to wait for nature to produce them all, and computer simulation provides a means of examining the possibilities. Results of this analysis indicate that a reliable estimate of potential loss cannot be determined solely from the density of population-at-risk in a region or from the meteorological characteristics of the area. The interaction of all four factors is needed.

For the hurricane hazard, the simulated production of a natural disaster can be illustrated by overlapping the severity patterns of wind and coastal flooding caused by a hurricane with the dense populations-at-risk along the Gulf and Atlantic coastlines.

Estimates of the variation in the catastrophe-producing potential of wind and coastal flooding along the coastline can be made by simulating the inland passage of a hurricane of identical physical characteristics from each of 100 landfall locations equally spaced from Texas to Maine. The results show that the potential loss varies greatly, both in terms of the size of the population-at-risk and of the magnitude of the resulting effect. The probability of occurrence also varies with landfall location.

A major problem associated with the use of natural hazard simulation is lack of information on the various populations-at-risk and on their vulnerability to natural hazard effects. Another problem is incomplete understanding of the physical phenomena associated with the event. The question, "How good is good enough?" must be asked continually.

Natural hazard simulation can provide unique types of information and insights into the workings of complicated situations, such as natural disasters. For example, we used this approach to produce thousands of "years" of simulated loss experience under various conditions for the U.S. Department of Housing and Urban Development. The results were helpful in developing the National Flood Insurance Program.[6]

Notes

1. Starr, C. 1969. Social benefit versus technological risk. Science 165:1232-38.

2. Boyd, D. W., *et al.* 1971. A decision analysis of hurricane modification. Stanford Research Institute, Menlo Park, Cal.

3. Reed, J. W. 1973. Cloud seeding at Rapid City: A dissenting view (with comments by E. Bollay; M. C. Williams; R. A. Schleusener & A. S. Dennis; P. St.-Amand, R. J. Davis & R. D. Elliott; and A. H. Murphy & S. W. Borland). Bull. Amer. Meteor. Soc. 54:676-84.

4. Reed, J. W. 1974. Another round over Rapid City (with comments by A. S. Dennis; R. D. Elliott; and A. H. Murphy & S. W. Borland). Bull. Am. Meteor. Soc. 55:786-90.

5. 197 Misc. 730, 97 N.Y.S. 2d 238 (Sup. Ct. 1950).

6. 42 U.S.C. secs. 4011 *et seq.* (1970).

Significance of the Conference:
A Scientist's Perspective

Thomas F. Malone, Director

Holcomb Research Institute
Butler University
Indianapolis, Indiana 46208

It would seem appropriate to recount here what was probably the first court case involving weather modification. During a severe drought in New York in the latter part of the nineteenth century, a minister named Duncan McLeod decided that there should be prayers for rain. The entire community, with the exception of Phineas Dodd, apparently agreed, and Duncan McLeod and his parishioners proceeded with prayers over his objections and did so with great fervor. Soon it was raining, but the rain was accompanied by unwanted lightning that struck Phineas Dodd's barn and destroyed it. Mr. Dodd sought recompense from Duncan McLeod, who refused to pay. During arguments at the trial, it appeared as though Phineas Dodd would win, but the judge—with the wisdom that we know judges to have—was persuaded to dismiss the case, observing that the rain was an answer to their prayers but that the lightning was a gratuitous gift of God.[1]

To put matters in current perspective, I remind you that our planet was launched about 5 billion years ago, that some kind of life appeared about 3 billion years ago, and that our atmosphere developed over the next billion years or so. Vertebrates appeared about 500 million years ago, with the dinosaurs arriving about 200 million years ago and lasting for about 140 million years. Precursors of man may have been here 4 or 5 million years ago, but modern man did not emerge until about 50,000 years ago. Now, if we are as smart as the dinosaurs, we should be able to last at least 100 million years, but to do so we must adjust our affairs to take into account this interval of time.

During the last several hundred years, we have learned to manipulate materials, transform energy, interfere with life processes, store and disseminate information, and otherwise change our evolutionary course. We have indeed arrived at a new level of consciousness or a new estate for mankind. However, we also are transforming our natural resource base into other forms of material and energy at an alarming rate, and we are doubling our population every few decades. Rather than attain even a greater level of consciousness, we could encounter a cascading series of disasters and never reach our 100 million year goal.

We must remember that our future will not be decided by a major plebiscite but in a manner described by John von Neumann in a timely article about 20 years ago entitled "Can We Survive Technology?"[2] After a few thousand carefully selected words, he finally answered his question with, "Yes, probably." He pointed out that mankind will face a long series of small decisions that in the aggregate will determine our destiny. Our task is to make these small decisions correctly and not to overlook their cumulative effect.

I think the scientists have succeeded at this conference in providing the lawyers with a glimpse into the world of meteorology and with a feeling of the zest with which atmospheric scientists try to solve these problems with such far-ranging implications for our civilization: the potential to reduce the 400,000 lives lost each year from weather-related disasters, to decrease the $6.6 billion property damage we bear annually, to enhance the beneficial use of natural resources, and to increase agricultural productivity. We can capitalize on the potential double-digit benefit/cost ratios.

We are all aware that the world faces a food crisis and that this country will be expected to continue to be the breadbasket for others. This awareness of the moral obligation to take care of our resources has developed during just the last one or two decades. The meteorologists share this awareness as they attempt to develop weather modification techniques that will help alleviate drought and otherwise increase the potential of our land and our atmospheric resources.

The conference also has provided the scientists an opportunity to observe, admire, and follow the closely reasoned line of thinking advanced by the lawyers. I am particularly impressed by their attitude of "We can do it, let's figure out a way," rather than the attitude so often associated with them of "You can't do this because it just won't comply with our laws and regulations." This forceful and positive approach is very heartening.

I also carry away the impression that the uncertainties concerning the workings of our societal mechanisms are increasing and that the scientists must work with the lawyers in reducing them and in overcoming mutual barriers to fulfilling their obligations toward society. I want to underscore what has been said several times about our inability to reduce to zero our uncertainty about the atmosphere. We must accept this large degree of uncertainty and learn to live with it. No matter how well we plan experiments, we will always face a residual uncertainty because no two clouds or thunderstorms or hurricanes are exactly alike. The question is, "Are we proceeding with the proper mixture of urgency and prudence?" We have heard several commentators say, "No," and I agree. However, we must remember, as also has been pointed out, that weather modification should be considered a means to an end. We need to look at benefit/cost ratios for alternative ways of achieving those ends and not focus solely on the ratios for weather modification. For example, if the national or international goal is to increase food production, we must compare the benefit/cost ratios for weather modification with those for experimentation with biological nitrogen fixation and for similar activities. Regardless of the promises of weather modification, we should constantly keep in mind the need to restrain our enthusiasm.

I perceive three clouds over this conference that have not been dissipated or made to precipitate. First, we must address the problems concerning institutional arrangements for weather modification, particularly within the federal government. Most of us agree that we can fashion more effective arrangements there and, without going into details, I simply would compare our current situation with the Global Atmospheric Research Program of the United Nations (GARP). This is far larger than a typical weather modification program and involves the launching of geosynchronous satellites by Israel, Russia, Japan, and the United States by 1978. You can see the organizational problems involved, but that program is proceeding on schedule.

I see a growing role—particularly concerning food production—for our land-grant universities. We need a lead agency to coordinate these activities—perhaps the National Oceanic and Atmospheric Administration—that will promote contracts or grants to these universities. If one agency undertakes the entire program, it will result in a loss of diversity and variety of approaches by individual researchers. This would be a superb opportunity to marshal the intellectual resources of this country around the pressing problems of weather modification. The universities have the potential, and we are devising completely new tools to assist in solving these problems. An example is the Doppler radar that allows one to peer inside a cloud and measure the three-dimensional field of motion. This is a fascinating breakthrough toward understanding the operation of these "rain factories." We also have the computing capacity and the models that no longer are so rudimentary. The time is right for mounting an effort of the magnitude commensurate with the benefits to be achieved.

Second, we have not discussed sufficiently the international aspects. I touched on this while describing GARP as a piece of the fabric that holds together this curious mixture of nations. Fortunately, the attitude at the international level is changing from one of reluctance toward one of enthusiasm about international cooperative efforts. The excitement within the World Meteorological Organization typifies this new sense of enthusiasm within the international scientific community.

Third, we must pay more attention to changes in public attitude over the next few years because of the growing recognition of inadvertent weather and climate modification. Two examples should suffice. We evidently are depleting the ozone concentrations in several ways that we do not thoroughly understand, which will result in an increase in the ultraviolet radiation that impinges on earth. Another example is the increase in carbon dioxide concentrations in the atmosphere. We now are trying to project these to the year 2175 and find, after making reasonable assumptions about the partitioning between atmosphere and ocean, that we might have as much as a five- to tenfold increase in carbon dioxide by then. Most of the carbon dioxide comes from combustion of fossil fuels and it acts as a shield to prevent infrared radiation from escaping to outer space. An increase in carbon dioxide acts to increase the temperature. Our admittedly crude models suggest that a doubling in carbon dioxide will raise the world temperature by about $2°C$, and calculations show that every subsequent doubling will result in another increase of $2°C$. So, it is not unreasonable to expect a worldwide increase in atmospheric temperature of $6°C$ in 200 years. This would be simply cataclysmic and would have profound implications for the world energy budget. I am not predicting a melting of polar ice caps but am reporting that thoughtful persons who are not prone to making rash statements are becoming increasingly concerned about the increase in atmospheric carbon dioxide.

All the problems discussed at this conference will be dealt with by a series of small decisions. We are making the first of these already and shall continue to do so for the foreseeable future. It will not be much longer before the magnitude of their effect will be placed in proper context by the public and its decision makers. Lawyers and scientists had best learn to work together. They will have no recourse but to do so as the importance of these decisions becomes more apparent.

Notes

1. Partridge, B. 1939. Country Lawyer (Whittlesey House, McGraw-Hill Book Co., New York) p. 77.
2. Neumann, J. von. 1955. Can We Survive Technology? Fortune, 51(6): 51-52, 106-08.

Significance of the Conference:
A Lawyer's Perspective
Milton Katz, Director

International Legal Studies
Harvard Law School
Cambridge, Massachusetts 02138

I am going to discuss the significance of weather modification as an illustrative case of the possibilities and problems of collaborative action among scientists, engineers, technologists, and lawyers. In a narrower and more operational sense, I will consider it as a test case of what the ABA-AAAS group may reasonably hope to accomplish.

My first theme relates to the many meanings of "facts," the many meanings of "fact-finding," and the many meanings of "fact reporting" in relation to the policies to be formulated, the programs to be elaborated, the decisions to be made, and the institutional settings within which the decisions are to be made. My second theme relates to the significance of the way in which the problem is defined and stated, again in relation to the policies to be formulated, the programs to be elaborated, the decisions to be made, and the institutional settings within which the decisions are to be made.

I begin by reminding you of an observation by Alfred North Whitehead in his *Adventures of Ideas*, to the effect that the movement of human society, viewed with historical rigor, reasonably may be described as the persistence of routines very slowly modified, illuminated by occasional short-range flashes of intelligence.[1] From my associations with scientists, which have been enormously profitable to me, I find that many of them instinctively feel that the legal profession and the legal system are the operational incarnations of stubborn routines and that the scientists represent the occasional short-range flashes of intelligence. However that may be, the relationship between the flashes of intelligence and the stubborn routines is a highly significant one. Hopefully, the flashes of intelligence may not only reveal the inadequacy of the routines but also prompt constructive changes in them.

Let me turn first to the many meanings of "facts," "fact-finding," and "fact reporting." I suggest that we keep these terms in quotation marks to remind us that typically we use each of them in a special sense of which we may not be aware. I take a "fact" to be the end result of a process or procedure that varies from profession to profession. Typically, "facts" are discussed in each profession without explicit recognition of the processes or procedures because the latter are taken for granted. For example, consider the differences between "facts" in the following contexts:

What does a "fact" mean to a chemist or a physicist?

What does a "fact" mean to an evolutionary biologist?

What does a "fact" mean to a paleontologist?

What does a "fact" mean to an archeologist?

What does a "fact" mean to a historian?

What does a "fact" mean to a competent newspaper reporter?

What does a "fact" mean to a legislator who is deciding whether to adopt certain legislation?

What does a "fact" mean to an agency administrator who is deciding whether to promulgate a regulation?

What does a "fact" mean to an agency administrator who is considering whether to prohibit a specific company from doing a particular thing?

What does a "fact" mean to a judge who is trying a criminal case that might result in deprivation of personal liberties?

What does a "fact" mean to a judge who is trying a civil case that might result in assessment of monetary damages?

It is clear that a different process is involved in each of these settings, with the end result being a "fact" for the purpose of reaching a decision within the particular institutional setting.

Compare what a question of "fact" means to an atmospheric physicist with what the same question of "fact" may mean to a commercial weather modifier who is seeking a commitment of new capital from an investment company. Is a "fact" for one seeking precise scientific knowledge the same as a "fact" sought as a valid basis for a commercial investment?

Dean Mann yesterday described his intensive study of the *Yuba City* case. He stated that the judge, who found that the plaintiffs had not satisfied the burden of proof in their case against commercial firms and the State of California, might have found that the plaintiffs carried the burden satisfactorily if the State of California had been the sole defendant.

It may be a fruitful line for the ABA-AAAS group of scientists and lawyers to explore and attempt to clarify the different interpretations of "fact" and "fact-finding." The lawyer and the scientist each reacts to terms as they are understood and used within the respective professions, but each may be unaware that the other understands the term differently. Even when aware of the difference, he may not fully appreciate its nature and significance. This perpetuates misunderstanding of more than a trivial nature.

In our current meeting, each of the working groups paid a great deal of attention to the need for dissemination of "facts." Each person who used that term knew what he or she meant and assumed that it was equally clear to the other persons present. To a detached observer, however, it was apparent that they were not talking about the same "facts" and therefore were not talking about the same processes of dissemination.

I turn now from the "fact-finding" process to the way in which problems are defined and stated. Lawyers are sensitive to the importance of how an issue is stated. Long experience has taught them that the mode of statement of an issue to a large degree controls the legal process that follows. Their experience and resulting approach seem to me highly relevant to the issues and problems discussed in our current meeting.

I will use one interchange in a working group earlier today to illustrate the point. A participant stressed that the current effort to increase the amount of precipitation is a limited enterprise. Few persons engage in it, he pointed out; the results influence only very limited geographical areas and a small number of people. In consequence, the participant questioned why other people make such a fuss about problems that really are not there. With the problem stated thus, the conclusion that we need not worry about societal implications seemed to follow very easily.

Someone else reminded the group that warnings and forebodings about pollution from automobile exhaust would not have registered in the early days of this century, when the number of cars in any basin might not have exceeded 50. In those circumstances, a notion that automobile pollution would alter the atmosphere to a significant degree would have appeared silly. With over 100 million cars on our roads, the problem acquires a radically different significance. We might want to project our thoughts to a time when a substantial number of commercial weather modifiers may be regularly at work throughout the country and then consider whether our judgments about the nature and significance of the problem might undergo a radical shift.

Tom Malone in his fascinating summary reminded us that we have learned relatively recently to care for the surface and atmospheric water on this planet: a vital resource in limited supply and in critically short supply in many areas. Tom insists we must take a holistic view to achieve optimal management of our total water resources.

We must look at the activities of any particular form of weather modification, such as precipitation augmentation, as an incidental part of the much larger problem of managing all the earth's resources. I reiterate what has been said earlier, both here and in the working groups, that we cannot limit our outlook to modification efforts intended solely for agricultural production. Our water resources sustain human health and our industrial economy, including the production of needed energy from sources now under contemplation. In addition, to many of us the prospect of a beautiful lake or river is a basic aspect of life. We cannot overlook the recreational and aesthetic uses of water. We must also take fresh account of the painful fact that many of our watercourses are used primarily to transport sewage and other wastes.

I return to the significance of how a problem is stated. What is meant by a "water shortage"? Shortage for what? In respect to which particular uses? The emphasis will vary with the predominant preoccupation of different regions at different times. Is the problem defined as the need to eliminate or reduce water quantities allocated to certain uses; or as a need to increase the supply; or as a need to design systems to increase the efficiency of a particular use?

Consider the interplay between "facts," "fact-finding," and the definition of the problem as typically understood by scientists and engineers and "facts," "fact-finding," and the definition of the problem as typically understood by lawyers, including practitioners, legislators, judges, and law teachers.

An undertaking to suppress hail or to increase precipitation involves an existing set of legal and governmental institutions. I use the word "institution" to mean both an organization and a concept, such as the concept of property. The weather modifier works in an area where people own land and will seek redress for any wrongs they feel they have suffered. The modifier also will work within a framework of legal doctrines, some of which may be used to the advantage of the modifier and some of which may be used to reduce or impede his activities. The law of our Western states gives effect to a doctrine of prior appropriation which, in simple terms, grants to the first person who makes use of water from a common supply the right to continue using that amount, regardless of what new developments upstream owners may have in mind. Due to plentiful rainfall and an inheritance from England, the law of our Eastern states gives effect to another doctrine called "reasonable use" under which riparian landowners are allowed to use the water of streams in a reasonable manner so that they do not unduly interfere with other users of the common supply. In a legal setting that accepts the notion of prior appropriation, it will be easier to persuade courts, legislatures, and administrative agencies to accept prior appropriation of clouds. The legal setting growing out of past evolution will affect future decisions.

What does this mean for us in this meeting? To the extent that scientists understand how problems are defined within the legal system, and how the definition of the problem relates to remedial procedures, the scientist can reduce his sense of exasperation and frustration at the legal system. He will also find it easier to adjust his own way of defining problems to maximize the possibilities of accommodation to the legal system while adhering to his own purposes. Similarly, to the extent that lawyers understand how scientists and engineers define problems and search for "facts," the lawyer may be able better to accommodate the operation of our legal and governmental institutions to scientific needs and opportunities. In doing so, both the scientists and the lawyers could help society take advantage of the benefits that science and engineering provide. Both also, of course, will be in a better position to foresee potential difficulties.

At this point, I would like to focus on a specific illustrative issue. Assume that a food processor wants to add a particular ingredient to a processed food. He is challenged by parties who want to prevent this. One party bases his challenge on the grounds that the ingredient might be carcinogenic. The other party bases his challenge on the grounds that the ingredient serves no useful purpose whatsoever but serves only as an excuse for a substantial increase in price by the processor. Now let's assume that scientists report that the ingredient might possibly be carcinogenic, but that present knowledge is inadequate to calculate the degree of risk. The question before the decider of fact, whether it be in the Food and Drug Administration or a court or a legislative committee, relates to the degree of risk. The risk may be stated in terms of medical risks that might arise from inclusion of the ingredient and economic risks and risks to freedom of enterprise that might be involved in a prohibition by the regulatory agency. The way in which the risks are stated will affect the evidence introduced. It can tip the whole decision-making process one way or another. In regard to the second challenge, involving a contention that the ingredient is a useless filler serving as an excuse for raising the price, the decision maker might respond that the customer can and should take care of himself by not purchasing the product. The unspoken setting of the competitive marketplace thus may define the issues and therefore which "facts" will be regarded as relevant to the decision.

The National Conference of Lawyers and Scientists should promote a greater awareness by decision makers of a need to examine carefully the questions, "How am I stating the problem?" and "Why do I state it this way?" and "How far does my statement reflect my professional habits or traditions that may not really be relevant or appropriate to this case?"

I would like to close by referring to the illuminating perspective given us by Tom Malone about our historical development. I will go back only to the seventeenth and eighteenth centuries, commonly taken as the period of origin of the scientific and technological (or industrial) revolutions. Given the dominant role of science and technology in the intellectual and practical life of our society ever since that period, isn't it remarkable that the need for greater cooperation among scientists, engineers, lawyers, and students of government should have surfaced as a subject of wide debate and concern so relatively recently? Of course, it isn't altogether new, as the development of administrative law and antitrust law attests. But despite the steady development of regulatory law in our society, it has not been accompanied by a corresponding development of an adequate mutual understanding among lawyers and scientists of the respective traditions and manners of approaching issues. This conference expresses the importance of a continuing effort to reach the kind of accommodations and understandings demanded by current and future events.

Note

1. Whitehead, A. N. 1933. Adventures of Ideas (Norton & Co., New York) p. 114-15.

Open Discussion

MODERATOR: Our two distinguished rapporteurs demonstrated once again that neither is only a lawyer or only a scientist. Both are staunch advocates of multidisciplinary undertakings, as demonstrated by worldwide recognition of their accomplishments.

THOMAS MALONE: I would like to emphasize, as Lou Battan did, that the only way to reduce uncertainty in atmospheric physics is to engage in field research. It simply cannot be done in the laboratory. It appears that the obviousness of this statement is not perceived within the federal bureaucracy, where a rather chaotic system inhibits precisely the kind of research needed to reduce the uncertainties. It results in lack of uniformity and a progression of regulatory complexities.

JAMES SMITH: As a federal employee, I am constantly aware that the state of California is not willing to accept an agency of the federal government as the lead agency in matters affecting the people of that state. We have some interstate problems, such as when we apply silver iodide in the Sierra Nevadas, but I seriously doubt whether a majority of the people in the West want the federal government to be given a supervisory capacity, even then.

MODERATOR: Dean Mann suggested that it would be a great idea for more law students and students of science to participate in each other's classes. My experience suggests that the institutional resistance to that is almost overwhelming in many universities. Talk about multidisciplinary cooperation increases, but the lack of results is discouraging. Science students appear more willing to come into the law schools than do the law students into the science schools, perhaps because the relevance of law to science seems more apparent than the importance of science to law.

CHARLES COOPER: Perhaps effective interactions between scientists and lawyers would be better if we considered the harmful effects of inadvertent weather modification. Just how much harm must be evident or how much potential harm must be anticipated before legal sanctions—either legislative or judicial—are possible to protect the health of people and the quality of our environment? The issues seem to be of far greater magnitude than those concerning intentional modification of weather.

WALLACE HOWELL: We have heard considerable comment about downwind effects of weather modification, perhaps even as far as 200 miles (320 km) from the target area. This really is inadvertent weather modification, and its effects need to be examined both by scientists and by lawyers.

It might not be justified to assume that all inadvertent effects will be harmful to the environment. In general, sterility is consonant with aridity and fecundity is consonant with moisture. When we look at inadvertent effects on a large scale, we very well might find that they are by and large beneficial.

CHARLES COOPER: That may be true of downwind phenomena, but I was referring primarily to atmospheric changes on an even larger scale, such as the increase in carbon dioxide concentration.

LEE LOEVINGER: I understand "inadvertent effects" to mean those that result from the overall activities of an increasingly technological civilization rather than the unintended and relatively incidental effects associated with deliberate attempts to modify weather. The effects of contemporary industrial cultures are by far the greater of the two, and we might want to clarify our terms.

NASH ROBERTS: Perhaps we should use the term "unintentional" rather than "inadvertent," because it does result from deliberate activities. Would it be appropriate to convene a subsequent conference to consider the scientific and legal uncertainties of unintentional weather modification?

HAROLD HORVITZ: We must remember that the purpose of this conference, or the next conference for that matter if there is one, is not to explore in depth the subject of weather modification, be it intentional or otherwise. Our purpose is to see how lawyers and scientists can get together to develop a methodology for resolving the uncertainties. This subject is only a means of addressing ourselves to these procedural and institutional problems.

JAMES SMITH: We are trying to cross-fertilize two disciplines that have not had much to do with one another in the past on a topic that is quite complex. We started by discussing precipitation augmentation for a particular use, moved on to hail suppression, and then on to diverting or dampening cyclonic storms. We have covered a great deal of scientific territory in a very short time, and I see nothing wrong with that if our purpose is to bring members of the legal and scientific professions together to consider mutual problems concerned with these uncertainties.

MODERATOR: The prime purpose of convening this conference is to increase communication and cooperation among lawyers and scientists, and we chose weather modification as a subject upon which to narrow the philosophical discussions because of the unique blend of scientific and legal questions surrounding it.

JAMES SMITH: Over the last three or four years, we have been inviting representatives of the legal profession to our annual Western Snow Conferences to talk about this very problem. I am convinced that it would be of value for members of the legal profession to make a greater effort to talk with scientists, not only about particular problems such as weather modification but also about the broader questions that impede or promote cooperative efforts. We are anxious to have lawyers speak at meteorological conferences or write for our journals, and I am confident that other scientific disciplines share this view. I visit my physician's office more than I do my lawyer's, so I really don't know what kind of magazines lawyers read, but I would hope that in the future they contain articles of more than a strictly legal nature.

MODERATOR: That's an interesting point. One of the first suggestions made after creation of the joint ABA-AAAS group was that it sponsor cross-publication of papers on legal and scientific issues in *Science* and in the *American Bar Association Journal* on a regular basis. I should say that *Science* does an extraordinary job of covering legal issues that are of interest to scientists, whereas I cannot say the same for the legal periodicals. There are a few exceptions—such as *Jurimetrics Journal,* published by the ABA Section of Science and Technology, and journals devoted to specialty topics, such as forensic science and computers—but very few of the legal periodicals with a broad readership do much to promote interdisciplinary understanding. This seems to be changing for the better in response to social realities rather than in response to perceived needs by most lawyers. The medical journals publish many articles and accounts on legal matters, but most of these are designed to satisfy special information needs, such as medical malpractice law.

EDWARD COLLINS: Many of the problems we face at the law-science interface are due to a lack of communication between lawyers and scientists in the past. Lawyers tend to cite a problem and then go into action without considering alternative solutions or technological possibilities. For example, the federal statutes concerning air and water pollution establish unrealistic goals, yet many lawyers seem to ignore the fact that the statutory standards cannot be met. Legislating a solution in the absence of technological capability is totally unrealistic. Laws might accelerate development of technology to satisfy their requirements, but we all would be far better off if the lawyers and scientists share information before enacting legislation of this sort.

MODERATOR: Several participants here can speak authoritatively about the federal or state legislative process.

EDWARD HELMINSKI: Yes, of course, members of these professions need to cooperate more actively. Let me go back a moment to the subject of indemnification and ask whether, in light of the experience with the Price-Anderson Act,[1] it would be wise to recommend such governmental indemnification of weather modification activities. The Price-Anderson Act indemnified the nuclear industry and, as a result, public support for increased nuclear development was weakened. The public increasingly has the attitude that if the nuclear industry cannot pay its own insurance premiums, then it simply is not safe enough for the public. If we have a governmental guarantee of indemnification for weather modification activities,[2] we might very well expect to find the same shift in public support. These indemnification programs raise genuine issues concerning science and technology to which the lawyers might not be giving sufficient attention.

LEE LOEVINGER: This illustrates the problem about the differences between private and public law. I think we can make a modest contribution to successful interchanges between scientists and lawyers at the private law level. That is, how lawyers can talk more confidently and effectively with scientists prior to litigation by learning more about handling of scientific data and this sort of thing. But how does anybody talk so simply to politicians? The Price-Anderson Act is strictly a political issue, and the lawyers and scientists involved are speaking as representatives or lobbyists of special interests or as politicians seeking election. They are lawyers and scientists who are advocates performing within the political arena. No doubt, lawyers and scientists have obligations to work for sound public laws and policies, but I don't know how we can encourage them to do so under our present system without becoming advocates or representatives of special interests. We need to encourage the participation of lawyers as lawyers and scientists as scientists.

HAROLD LEVENTHAL: I am having difficulty generalizing from the particular points we are mentioning in accordance with good inductive reasoning, and I stub my mental toes when trying to put this stimulating discussion into more general categories.

Lawyers serve society in two general manners, in the adversary setting and in the constructive setting. The former is familiar to all of us—from television shows, if not otherwise—and scientists quite properly are wary of lawyers in this adversary situation. Scientists will become dissatisfied with the cross-examination procedures and with the difficulties of educating their own attorney. Anything that we can do to lessen the mutual difficulties during these processes would be beneficial. It is difficult for scientifically trained individuals to be comfortable when lawyers ask questions in what might appear to be unfair terms and attack a witness's integrity. However, I don't know how to develop any better code of courtesy while retaining the advantages of the adversary system. We are experiencing some changes in this system, as noted recently by Judge Frankel,[3] as we put more emphasis on a "search for truth" rather than on the procedural matters about who is to joust with whom, but these changes often are glacially slow.

On the constructive side, all businessmen are accustomed to using lawyers to set up new enterprises, and governmental agencies rely on lawyers to work out a myriad of institutional arrangements.

Dean Acheson in his *Present at the Creation*[4] recounted the dismay of

Lord Keynes at the number of lawyers brought to Bretton Woods by the Americans during their negotiations. The British may have had one or two lawyers, whereas the Americans came with a full complement of them. However, at the conclusion of the conference, Lord Keynes acknowledged to Dean Acheson his appreciation for the lawyers' presence because they had helped sharpen the issues and had drafted a charter that would be much more flexible than otherwise would have been the case.

Lawyers bring into any constructive process their familiarity with structure and process, while leaving the substantive matters to the scientists. The scientists provide the real determinants of policy and the lawyers provide the frameworks within which they operate.

We heard earlier about a suggested commission to investigate the real and foreseeable problems with weather modification, but I don't quite see how the lawyers and scientists will be expected to cooperate without an adversary problem or a constructive problem upon which to focus. For this to become a functional reality, the members must be given some specific goals toward which they can apply their respective talents, and some of these appear to be adversarial and some constructive.

EMILIO DADDARIO: If I believed that we can't talk to the policy makers—the politicians—and that we as a result must rely upon the power centers and the lobbyists, I would suggest that we save our time and go home. I just don't believe this to be the case and I think that in certain areas we have communicated quite well.

Our main communications problems are in the contentious areas where the unknowns are the greatest or where the evidence suggests that the negative aspects of action are greater than the positive. I agree that we must increase cooperation to promote development of the positive aspects. It is possible that the Congress in the 1960s could have laid a stronger framework for the air and water pollution legislation. The Congress did hold some hearings on technical adequacy to meet the problems of pollution and then concluded that we should examine available technology, regulate to the best of our ability within the confines of that technology, and keep abreast of developments over the years as technology and our regulatory capabilities improve.

Then, the admission by one of the automobile manufacturers that it had hired detectives to trail Ralph Nader heightened the public's emotion and affected the ability of Congress to deal in a more rational way with the problems. What started out as a good faith effort suddenly went astray.

However, let me provide several examples of why I think we can make the necessary positive gains. When President Nixon disbanded the Office of Science and Technology in the White House, the protest did not come primarily from the scientific community but from the Congress. We can take some hope because the Congress, based on hearings held some years earlier and on ones held at that time, almost immediately began to reconsider the re-establishment of this activity, with the unmistakable desire to have a science decision-making presence in the White House to assist in setting policy at the very top level.

Congress was competent to consider these issues because of its experience with restructuring the National Science Foundation and with fulfilling the national needs for scientific and technical funding and manpower, even during times when these programs were not in favor with the administration. Before closing, I would like to quote briefly from the first section of the bill before the U.S. House of Representatives that would establish a science and technology policy for the United States and reintroduce a science advisor in the White House:

> The Congress, recognizing the profound impact of science and technology on society, and the interrelations of scientific, technological, economic, social, political, and institutional factors, hereby finds and declares that (1) the general welfare, the security, the economic health and stability of the Nation, the conservation and efficient utilization of its natural and human resources, and the effective functioning of government and society require vigorous, perceptive support and employment of science and technology in achieving national objectives; (2) the many large and complex scientific and technological factors which increasingly influence the course of national and international events require appropriate provision, involving long-range inclusive planning as well as more immediate program development, to incorporate scientific and technological knowledge in the national decisionmaking process; (3) the scientific and technological capabilities of the United States, when properly fostered, applied, and directed, can effectively assist in improving the quality of life, in anticipating and resolving critical and emerging international, national, and local problems, in strengthening the Nation's international economic position and in furthering its foreign policy objectives. . . .[5]

I believe that the situation is much more encouraging than some people here believe and that it is conducive to responsible attempts to introduce the best available knowledge into the legislative process. I expect in the future to see greater contributions by scientists and lawyers working together in federal legislative and executive activities. And in our own manner, I believe that the ABA-AAAS group can assist materially by encouraging these activities and by directing them toward where they will be of greatest national benefit.

MODERATOR: That sounds to me very much like a closing statement and I suggest we adjourn on that note.

Notes

1. 42 U.S.C. sec. 2210 (Supp. V 1975).

2. Indemnification of weather modifiers by states after "extraordinary operations" was proposed by F. Maloney, R. Ausness & J. Morris, A Model Water Code, With Commentary 345-49 (1972), and criticized by Thomas, Book Review, 3 Ecol. Law Q. 885 at 891-93 (1973).

3. Frankel, M. E. The search for truth: an umpireal view, 123 U. Pa. L. Rev. 1031 (1975).

4. Acheson, D. 1969. Present at the Creation (Norton & Co., New York).

5. Enacted as the National Science and Technology Policy, Organization, and Priorities Act of 1976, 42 U.S.C. sec. 6601 (Supp. VI 1976).

Conference Participants

(*Indicates persons unable to attend at the last minute)

Frederick R. Anderson, Esq.
Exec. Director
Environmental Law Institute
Du Pont Circle Bldg.
Washington, D.C. 20036

*Dr. David Atlas
Nat'l. Ctr. for Atmospheric Research
Boulder, CO 80303

*Prof. Michael S. Baram
Room 48-335
Massachusetts Institute of Technology
Cambridge, MA 02139

Prof. Louis J. Battan, Director
Institute of Atmospheric Physics
Univ. of Arizona
Tucson, AZ 85721

*Hon. David L. Bazelon, Chief Judge
U. S. Court of Appeals
Washington, D.C. 20001

Haradon Beatty, Esq.
Holland & Hart
500 Equitable Bldg.
Denver, CO

Prof. William Bevan
Dept. of Psychology
Duke Univ.
Durham, N.C. 27706

*Dr. Richard H. Bolt
Bolt, Beranek & Newman, Inc.
50 Moulton St.
Cambridge, MA 02139

*Dr. D. Ray Booker, President
Aeromet, Inc.
Box FF
Norman, OK 73069

Dr. Stewart W. Borland
Economics Branch
Agriculture Canada
Sir John Carling Bldg.
Ottawa, Ontario K1A 0C5

Prof. Roscoe R. Braham, Jr.
Dept. of Geophysical Sciences
Univ. of Chicago
Chicago, IL 60637

Prof. Daniel A. Bronstein
Natural Resources Bldg.
Michigan State Univ.
East Lansing, MI 48824

*Mr. William D. Carey, Exec. Officer
American Assoc. for the
   Advancement of Science
1776 Massachusetts Ave., N.W.
Washington, D.C. 20036

Mr. John T. Carr, Jr., Director
Weather Modification and Technology
   Division
Texas Water Development Board
Stephen F. Austin State Office Bldg.
Austin, TX 78711

Dr. Stanley A. Changnon, Head
Atmospheric Sciences Section
Illinois State Water Survey
Box 232
Urbana, IL 61801

William Clayton, Esq.
U.S. Attorney for District of
    South Dakota
U.S. Courthouse
Sioux Falls, S.D. 57102

Dr. Frederic N. Cleaveland, Provost
Duke Univ.
Durham, N.C. 27706

Edward J. Collins, Esq.
Iandiorio & Collins
60 Hickory Dr.
Waltham, MA 02154

Prof. Charles F. Cooper
Ctr. for Regional Environmental
    Studies
San Diego State Univ.
San Diego, CA 92182

Dr. James W. Curlin
Oceans and Coastal Resources Project
Congressional Research Service
Library of Congress
Washington, D.C. 20540

Mr. Emilio Q. Daddario, Director
Office of Technology Assessment
Congress of the United States
Washington, D.C. 20510

Prof. Ray J. Davis
School of Law
Univ. of Arizona
Tucson, AZ 95721

Dr. Ruth M. Davis
Institute for Computer Sciences
    and Technology
Nat'l. Bureau of Standards
Washington, D.C. 20234

Dr. Arnett Dennis
Institute of Atmospheric Science
South Dakota School of Mines &
    Technology
Rapid City, S.D. 57701

Dr. Vincent P. Dole
Rockefeller University Hospital
New York, N.Y. 10021

Dr. Earl G. Droessler
Administrative Dean for
    University Research
North Carolina State Univ.
Raleigh, N.C. 27607

Dr. Barbara C. Farhar
Human Ecology Research Services
855 Broadway
Boulder, CO 80302

Dr. John W. Firor, Exec. Director
Nat'l. Ctr. for Atmospheric Research
Boulder, CO 80303

Donald W. Frenzon, Esq.
Assistant General Counsel
Nat'l. Science Foundation
Washington, D.C. 20550

Dr. Don A. Friedman
Assoc. Director of Research
The Travelers Insurance Co.
One Tower Square
Hartford, CN 06115

Mr. William T. Golden
Treasurer, American Assoc. for the
    Advancement of Science
40 Wall Street
New York, N.Y. 10005

Prof. Lewis O. Grant
Dept. of Atmospheric Science
Colorado State Univ.
Fort Collins, CO 80523

Prof. Harold P. Green
Nat'l. Law Center
George Washington Univ.
Washington, D.C. 20052

*Dr. J. Eugene Haas
Institute of Behavioral Science
Univ. of Colorado
Boulder, CO 80302

Roger P. Hansen, Esq.
Hansen & O'Connor
American Nat'l. Bank Bldg.
Denver, CO 80202

Dr. Edward L. Helminski
Energy Program
Nat'l. Governors' Conference
1150 17th Street, N.W.
Washington, D.C. 20036

Haywood H. Hillyer, Jr., Esq.
Milling, Benson, Woodward,
   Hillyer & Pierson
Whitney Bldg.
New Orleans, LA 70130

Prof. Marcus E. Hobbs
Dept. of Chemistry
Duke Univ.
Durham, N.C. 27706

Harold Horvitz, Esq.
Guterman, Horvitz, Rubin & Rudman
Three Center Plaza
Boston, MA 02108

Dr. Wallace Howell
Bureau of Reclamation
Mail Code 1200
Denver Federal Center
Denver, CO 80225

Dr. William Jolly
Environmental Policy Div.
Congressional Research Service
Library of Congress
Washington, D.C. 20540

Helene C. Kaplan, Esq.
Emil, Kobrin, Klein & Garbus
540 Madison Ave.
New York, N.Y. 10022

Prof. Milton Katz, Director
Internat'l. Legal Studies
Harvard Law School
Cambridge, MA 02138

Dr. Conrad G. Keyes, Jr.
Exec. Secretary
No. American Interstate Weather
   Modification Council
New Mexico State Univ.
Las Cruces, N.M. 88003

Arthur F. Konopka, Esq.
Law, Science & Technology Program
Nat'l. Science Foundation
Washington, D.C. 20550

J. Charles Kruse, Chief
Tort Section
Dept. of Justice
Washington, D.C. 20025

Mr. Noel R. LaSeur, Director
Nat'l. Hurricane & Experimental
  Meteorology Laboratory
Univ. of Miami Computing Center
Coral Gables, FL 33124

John D. Lane, Esq.
Hedrick & Lane
1211 Connecticut Ave., N.W.
Washington, D.C. 20036

Mr. Henry Lansford
Nat'l. Ctr. for Atmospheric Research
Boulder, CO 80303

Mr. R. L. Lavoie
Environmental Modification Office
Nat'l. Oceanic & Atmospheric Admin.
Rockville, MD 20850

Prof. Victor J. Law
School of Engineering
Tulane Univ.
New Orleans, LA 70118

Hon. Harold Leventhal
Circuit Judge
U.S. Court of Appeals
Washington, D.C. 20001

Dr. Harold Lewis
Dean of Faculty
Duke Univ.
Durham, N.C. 27706

Lee Loevinger, Esq.
Hogan & Hartson
815 Connecticut Ave.
Washington, D.C. 20006

Dr. John McKinney, Dean
Graduate School
Duke Univ.
Durham, N.C. 27706

Lawrence G. Mallon, Esq.
Ass't. Sea Grant Director
4676 Admiralty Way
Marina Del Ray, CA 90201

Dr. Thomas F. Malone, Director
Holcomb Research Institute
Butler Univ.
Indianapolis, IN 46208

Prof. Dean E. Mann
Dept. of Political Science
Univ. of California,
Santa Barbara, CA 93106

Prof. Howard Margolis
Ctr. for Internat'l. Studies
Massachusetts Institute of Technology
Cambridge, MA 02139

Ronald A. May, Esq.
Wright, Lindsey & Jennings
2200 Worthen Bank Bldg.
Little Rock, AK 72201

Dr. Louis H. Mayo, Director
Program of Policy Studies in
  Science & Technology
George Washington Univ.
Washington, D.C. 20052

Dr. Alan D. Morris
Morris & Ward
Consulting Engineers
P.O. Box 5937
Washington, D.C. 20014

W. Brown Morton, Jr., Esq.
Morton, Bernard, Brown,
  Roberts & Sutherland
1700 K Street, N.W.
Washington, D.C. 20006

Prof. J. D. Nyhart
Coordinator of Law-Related Studies
Office of the Chancellor
Massachusetts Institute of Technology
Cambridge, MA 02139

Prof. George W. Pearsall
Dept. of Mechanical Engineering
Duke Univ.
Durham, N.C. 27706

Prof. Kenneth Pye, Dean
School of Law, Duke Univ.
Durham, N.C. 27706

Mr. Nash C. Roberts
521 Royal Street
New Orleans, LA 70130

Prof. David J. Rose
Dept. of Nuclear Engineering
Massachusetts Institute of Technology
Cambridge, MA 02139

Dr. Richard A. Scribner
American Assoc. for the
    Advancement of Science
1776 Massachusetts Ave., N.W.
Washington, D.C. 20036

Prof. Melvin G. Shimm
School of Law, Duke Univ.
Durham, N.C. 27706

Dr. James Smith
Pacific Southwest Forest & Range
    Experiment Station
U.S. Forest Service
Berkeley, CA 94720

Dr. Kenneth C. Spengler
Exec. Director
American Meteorological Society
45 Beacon Street
Boston, MA 02108

Dr. Pat Squires
Desert Research Institute
Univ. of Nevada
Reno, NV 89507

Elizabeth Stein, Esq.
American Bar Association
1155 E. 60th Street
Chicago, IL 60637

*Mr. Preble Stolz, Director
Office of Planning and Research
Office of the Governor
State of California
Sacramento, CA 95814

Prof. Howard J. Taubenfeld
School of Law
Southern Methodist Univ.
Dallas, TX 72575

William A. Thomas, Esq.
American Bar Foundation
1155 E. 60th Street
Chicago, IL 60637

Thomas J. Whalen, Esq.
Condon & Forsyth
1251 Avenue of the Americas
New York, N.Y. 10020

Hon. H. Emory Widener, Jr.
Circuit Judge
U.S. Court of Appeals
Abingdon, VA 24210

*Gary L. Widman, Esq.
General Counsel
Council on Environmental Quality
722 Jackson Pl., N.W.
Washington, D.C. 20006

Dr. Merlin Williams, Act. Director
Weather Modification Program Office
Nat'l. Oceanic & Atmospheric Admin.
Boulder, CO 80302

Mr. Joseph Wisniewski
Dept. of Environmental Studies
Univ. of Virginia
Charlottesville, VA 29903

Dr. Charles P. Wolf
Office of Technology Assessment
Congress of the United States
Washington, D.C. 20510

*Prof. James W. Zirkle
Law Center, Univ. of Mississippi
Oxford, MS 38677

Index